Portfolio Assessment of the Department of State Internet Freedom Program

Ryan Henry, Stacie L. Pettyjohn, Erin York

Prepared for U.S. Department of State, Bureau
of Democracy, Human Rights, and Labor

For more information on this publication, visit www.rand.org/t/rr794

Library of Congress Cataloging-in-Publication Data is available for this publication.
ISBN: 978-0-8330-8769-0

Published by the RAND Corporation, Santa Monica, Calif.

Preface

As a growing number of states take steps to censor, monitor, and control the Internet, the U.S. Department of State's Bureau of Democracy, Human Rights, and Labor (DRL) has sought to counter these efforts and protect Internet freedom. The purpose of this research is to assess the performance, balance, synergy, risk, and cost of DRL's portfolio of Internet freedom projects during fiscal year 2012–2013.

This research should be of interest to policymakers, activists, and analysts concerned with the United States' Internet freedom agenda and the connection between new communications technologies and democratization. This works compliments previous RAND research on Internet freedom, in particular the 2013 report *Internet Freedom and Political Space*, which explored in detail the links between Internet freedom and popular mobilization. This research was sponsored by DRL and conducted within the International Security and Defense Policy Center of the RAND National Security Research Division under contract number S-LMAQM-11-GR-585. The National Security Research Division conducts research and analysis on defense and national security topics for the U.S. and allied defense, foreign policy, homeland security, and intelligence communities and foundations and for other nongovernmental organizations that support defense and national security analysis.

For more information on the International Security and Defense Policy Center, see http://www.rand.org/nsrd/ndri/centers/isdp.html or contact the director (contact information is provided on the web page).

Comments or questions on this report should be addressed to the project leader, Ryan Henry, at rhenry@rand.org.

Contents

Figures

Summary

The United States has long argued that all people have a fundamental right to freely express and share their ideas.[1] Because the Internet enables individuals to communicate independent of time or distance, it facilitates the free flow of information. Yet some repressive states attempt to limit the content available online and use the Internet to identify and track those who oppose their rule. The U.S. Department of State's Bureau of Democracy, Human Rights, and Labor (DRL) Internet freedom program seeks to counter the efforts of authoritarian regimes to censor, monitor, and control the Internet.[2] DRL sponsored RAND in an assessment of the Internet freedom program's portfolio of 18 projects funded in fiscal year 2012–2013 to determine the program's effectiveness in portfolio performance, balance, and synergy among projects.

Findings

The primary goals of the portfolio assessment were to find out if DRL is effectively managing its Internet freedom portfolio and implement-

[1] Elizabeth Dickinson, "Internet Freedom: The Prepared Text of U.S. Secretary of State Hillary Rodham Clinton's Speech, Delivered at the Newseum in Washington, D.C.," *Foreign Policy,* January 21, 2010.

[2] Daniel Baer, Deputy Assistant Secretary, DRL, "Promises We Keep Online: Internet Freedom in the Organisation for Security and Cooperation in Europe (OSCE) Region," Statement Before the Commission on Security and Cooperation in Europe (U.S. Helsinki Commission), Washington, D.C., July 15, 2011.

ing its stated strategy. In addition, we wanted to know the expected value of the portfolio and how is it performing.

Performance

We based the Internet freedom portfolio assessment on the cumulative performance value of individual projects within the portfolio. The Internet freedom performance value of the individual projects was determined by each project's overall contributions to four major variables that affect Internet freedom and political space: (1) access to the Internet, (2) anonymity and security when online, (3) awareness and understanding of security threats and protective measures, and (4) advocacy for ensuring a free and open Internet compatible with the aforementioned U.S. policy toward free speech and human rights online. A project's risk level and cost, both direct and indirect, to the U.S. government (USG) also influenced its portfolio contribution. Under this methodology, the highest-performing projects were active in multiple areas of Internet freedom at comparatively low risk and cost.

To determine the portfolio analysis method that we used to calculate performance, we used DRL's stated strategy because a key objective of our analysis was to determine if DRL was implementing its desired strategy. Therefore, our analysis does not get at fundamental questions about DRL's underlying strategy. Yet this approach does have the virtue of forcing DRL to explicitly state and discuss its strategy, and allows one to evaluate whether the declared strategy is being put into practice in DRL's Internet freedom portfolio.

During fiscal year 2012–2013, DRL's strategy to enhance Internet freedom consisted of the following four major components:

1. countering online content restrictions by developing anticensorship technologies and expanding access to information and communications platforms
2. developing secure communications technologies to strengthen privacy and anonymity online
3. teaching individuals about good digital safety practices through training
4. advocating for national and international policies that protect Internet freedom.

Although all of these elements are important, DRL's strategy prioritized efforts to enhance anonymity and enable secure communication. Moreover, while access and awareness were equally valued, advocacy was considered to be less critical and, therefore, did not contribute as much to a project's overall performance.[3]

Consequently, projects that focused primarily on advocacy had relatively low performance scores, and projects that made a key contribution to anonymity on the Internet were among the higher performers. DRL's strategy aimed to correct what was thought to be an overemphasis in previous solicitations on circumvention or access without considering privacy. Increasingly, it was recognized that providing access without anonymity could endanger those who employ circumvention technologies, especially if they are unaware of their vulnerabilities while using these tools.

In a broad sense, this assessment of the overall portfolio found a strong diversity of effort and balance across the four Internet freedom variables. Each of the four performance aspects was addressed by multiple projects, a dispersion that adds robustness to the portfolio, given the challenges in measuring direct effects in this area. Additionally, there was a clear connection between DRL's stated program objectives and the projects' objectives and lines of effort.

Balance

The performance assessment revealed that DRL's Internet freedom portfolio was balanced with respect to project focus and geographical distribution. The quantitative assessment found that projects cluster into five functional types: technology development, training, technology testing, advocacy, and mixed efforts.[4] The portfolio also incorporates a mix of

---

[3] The DRL program strategy was communicated to the RAND team in a series of meetings in which DRL proposed and approved the value weighting discussed in Chapter Two.

[4] There were five functional types of projects: (1) programs that concentrated on developing new circumvention or anonymity technologies (which largely correlates with access and anonymity); (2) programs that focused on training at-risk populations to improve their understanding of online vulnerabilities and good security practices (which largely correlates with awareness); (3) advocacy projects that aim to support Internet freedom within states and international organizations; (4) test programs that worked toward ensuring that Inter-

high-risk and high-gain projects, along with some implementers who use more established approaches. All funded projects were clearly aligned with Internet freedom objectives, but the projects varied substantially on approach, breadth, geopolitical focus, and investment allocation.

Although the Tor project (also known as The Onion Router)—an anonymity technology based on proxy routing—is not a direct DRL grantee, several projects in the DRL portfolio employed it. Nine of the DRL projects used Tor to some extent, with most deriving a benefit in access and anonymity.

Synergy

A key finding from the portfolio assessment was that the program's total effect is greatly enhanced by the interaction and collaboration between implementers. Projects from the five cluster areas above intersect within the portfolio and produce opportunities for project synergy that can lead to enhanced project, as well as portfolio, performance or additional collaboration beyond the scope and timeframe of the DRL grant. The potential benefits of this element are substantial, as illustrated in Figure S.1. For example, projects engaging in technology development will benefit from interaction with groups testing that technology for security flaws. In addition, training programs may distribute newly developed circumvention tools.

RAND researchers mapped the existing relationships among DRL's projects to identify the level of synergy in the Internet freedom portfolio.[5] This analysis revealed that many projects informally cooperated or formally collaborated with other DRL grantees, but these

---

net freedom technologies were robust and did not have security flaws that put their users at risk (these programs did not fall neatly into one of the four Internet freedom variables); and (5) mixed programs that were multifaceted and had elements of all of the above.

[5] Ideally, one would want to foster collaboration across USG-funded Internet freedom projects, which would include those who receive grants from the Broadcasting Board of Governors (BBG), DRL, Defense Advanced Research Projects Agency (DARPA), and United States Agency for International Development (USAID). But systematically and comprehensively assessing crossdepartmental Internet freedom synergy was beyond the scope of this project, which was focused only on DRL. Moreover, because interagency cooperation and coordination is often quite difficult, we focused on assessing synergy among the limited group of grantees that DRL can directly influence—those within its own portfolio. Nev-

Figure S.1
Idealized Connections Among DRL's Internet Freedom Projects

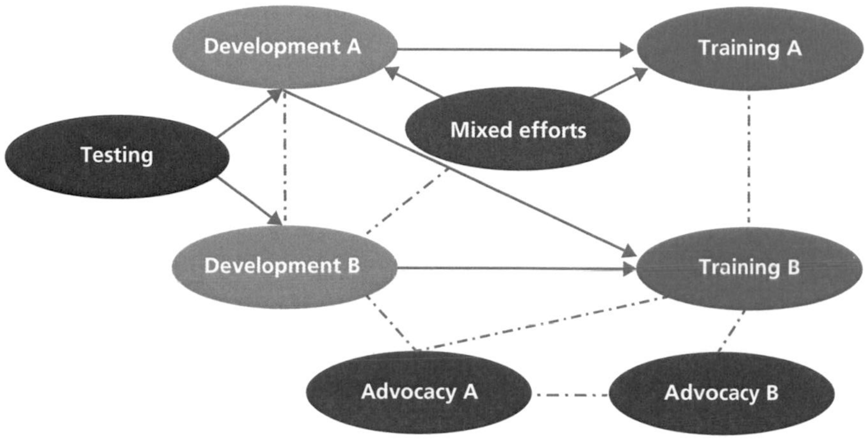

RAND *RR794-S.1*

connections were largely ad hoc and based on preexisting personal and professional relationships. We also found that several DRL projects used Tor, which produced an added element of synergy. While there are some connections among DRL grantees, more networking would be valuable in realizing the portfolio's full synergy potential. Therefore, to encourage its grantees to establish mutually beneficial connections, DRL should continue to hold regular implementers' meetings and encourage the use of established trusted communications channels.

There is an inherent tension between cooperation and competition in a limited-resource environment, but DRL could address this challenge in its selection criteria. By making formal or informal cooperation an explicit standard by which project proposals are judged, DRL could incentivize additional collaboration. Synergy presents an especially worthwhile investment because of its low cost and high potential payoff. Even if projects do not immediately collaborate, the latent relationships facilitated by DRL may produce lasting value. In particular, by fostering personal and organizational ties and enhancing

---

ertheless, promoting connections among the broader USG Internet freedom community merits further study and consideration.

trust, the Internet freedom community would be positioned to organically and swiftly respond to rapidly developing crises associated with Internet freedom.

Additional Observations

Technology

Developing new technologies that enable individuals to have unfettered and secure access to the Internet is a significant but complicated part of the DRL portfolio. The struggle between those promoting Internet freedom and those trying to control and monitor the Internet is a fast-paced game of cat and mouse. Consequently, the speed of this contest often outstrips the grant cycle, and implementers often have to modify their proposed deliverables in response to developments on the ground. Responding effectively to the countermoves made by authoritarian governments is difficult under any circumstances. This predicament is further complicated by the fact that technology development is not a traditional State Department activity and, therefore, not one of its core capabilities. Partnering with other USG entities that have proven technology development infrastructure could help overcome this limitation.

While developing technology—through both evolutionary improvements to existing circumvention and anonymity tools and the incubation of new revolutionary Internet freedom capabilities—is a critical component of DRL's Internet freedom portfolio, it alone cannot produce a free and open Internet. There is not a purely technical solution that would guarantee Internet freedom. Instead, it is a political struggle that takes places in many arenas, including legislatures, courts, and international organizations, and therefore requires a multifaceted response.

Enduring Value of Portfolio

DRL's investment in Internet freedom should have enduring value that extends well beyond the life of the individual grants. In particular, one of the most important effects of the DRL portfolio appears to be the community it nurtures. Although more might be done, DRL has taken steps to maximize the return on its Internet freedom portfolio by bringing together the individuals, organizations, and tools it supports. The DRL Internet freedom community could play different roles at different times. Typically, the DRL Internet freedom community is focused on increasing the State Department's steady-state capability to promote freedom online by encouraging formal and informal collaboration between grantees to improve the efficacy of both projects. At the same time, fostering these ties also develops a latent surge capacity to respond during Internet freedom–associated crises. At critical moments, the DRL-sponsored community has the capacity to rapidly and independently respond to developments in an effort to expand political space. In these circumstances, a self-synchronizing community composed of independent actors whose interests are aligned with the USG is not only well-positioned to react, given its grassroots connections, but it is also more agile and capable of responding in a timely fashion than the government. Moreover, having the USG stay in the background reduces the potential for blowback.

Risk and Cost

In addition to assessing the DRL portfolio's balance, we assessed potential risk—defined as the probability that a project would be successfully implemented—and cost. We evaluated the following three types of risk:

1. capability, or the soundness of a project's approach and staffing
2. acceptance, or the likelihood that intended users would willingly adopt the project's offering
3. sustainability, or the probability that the project would be able to carry on beyond the term of the DRL grant.

We then assessed individual projects against these risks to assess their effect on the cumulative portfolio risk. In short, we found that DRL's portfolio is characterized by an acceptable amount of risk tolerance.

One area where the portfolio appeared to accept risk was in technology development. DRL's portfolio included apparently low- and high-risk technologies for development, which correlated closely with the evolutionary or revolutionary nature of the technology under development. We viewed DRL's failure-tolerant approach as both healthy for the technology development community and important for advancing Internet freedom capabilities.

To assess the costs associated with DRL's Internet freedom portfolio, we used several components, including direct costs (the project's level of funding) and indirect costs (such as domestic or international political costs). We found that direct costs were relatively evenly divided among the grantees, with the better-resourced projects generally focusing on multiple aspects of Internet freedom. Political cost assessed the likelihood that a project might generate negative diplomatic, domestic political, or media effects. Across the DRL suite of projects, we found the assessed political costs had yet to materialize, which does not mean that problems might not arise in the future.

Project Execution

RAND researchers found several key components that characterized well-run projects; principal among these components were a staff that includes a visionary or idea champion, a skilled functional specialist, and a competent program manager. At times, one person might be responsible for all of these tasks, but more often, projects had a different person filling each role. Another key component among the best projects was healthy interactions with other Internet freedom programs and the larger Internet freedom community.

Common Challenges

During the assessment, we found that many DRL implementers face shared challenges. One of the most common difficulties encountered by grantees was securing and retaining skilled technologists at a nonprofit organization's salary. Given that there are much more lucrative careers in the private sector, implementers had to search for qualified personnel with technical skills who were primarily motivated by the cause of advancing Internet freedom. While such dedicated tech-savvy individuals do, fortunately, exist, their numbers are few and their talents highly sought.

Many of the projects were based on a compelling idea and dedicated staff, but not all projects had a sufficient cadre of experienced developers and managers to ensure their long-term success. Some lacked experience in negotiating the multistakeholder environment that characterizes the Internet freedom governance model.

Another issue many DRL implementers faced was finding ways to deal with rapidly changing circumstances or crises. On some occasions, real-time developments raised new issues or challenges not addressed in the original grant. Implementers, therefore, desired more flexibility so that they could modify their activities to respond to unforeseen opportunities and provide greater value.

Beyond Tor, one potential challenge area yet to be negotiated by any of the projects in the DRL portfolio is their ability to scale beyond pilot-project demonstrations. There are common growth hurdles experienced by both for-profit and not-for-profit organizations that need to be overcome if any of these projects are to reach beyond niche impact.

Internet Freedom and an Enlarged Political Space

As part of a more theoretical research effort, RAND researchers found that there is a positive, but indirect, relationship between Internet freedom and the expansion of political space in societies. Historically, there is an observable relationship between online and offline mobilization. The Internet has played an important role in expanding political space by eliminating distance and time as constraints to sharing information and building trust networks. Nevertheless, Internet freedom has not typically been a primary causal factor that directly increases political space. It has enabled the creation of new social movements, rather than been the primary factor in their formation. In other words, the Internet facilitates the creation and expansion of decentralized and broad social movements by increasing the availability of information, fostering the formation of a broad collective identity, and reducing the costs of collective action, but there also must be underlying grievances that galvanize a population. Moreover, the effect of Internet freedom on political space depends on the regime's repressive capacity. Of course, it is also important to note that the Internet has not had a uniformly positive effect; it has also enhanced the ability of states to monitor their opponents and spread disinformation and propaganda. DRL's projects seek to ensure that all people have free access to the Internet, which could be a critical enabler that helps to empower opposition movements within repressive states.

Generally, there are two major trade-offs an Internet freedom policy needs to balance: deepening versus broadening, and steady state versus crisis. First, an Internet freedom strategy needs to ensure it targets both opinion leaders and the broader population. In other words, it needs to deepen by focusing on those who already desire Internet freedom (opinion leaders), but also broaden by drawing in the general populace to increase the number of people who are mobilized and politically active. Second, an Internet freedom strategy needs to find the appropriate balance between increasing individuals' ability to securely access that Internet every day and intervening during crises when Internet freedom could potentially tip the balance toward greater political space.

Internet Freedom as a Cost-Imposing Strategy

While Internet freedom is consistent with the United States' ideological and economic interests, somewhat unusually, it also largely aligns with the country's national security interests.[6] Equally important, by investing relatively little in Internet freedom initiatives (approximately $30 million a year to the Department of State, USAID, and BBG), the United States can impose costs on authoritarian rivals who are forced to devote significantly more resources to maintaining domestic stability.[7] Cost imposition is a not a stated objective of DRL's Internet freedom program, but it is an ancillary benefit that could be particularly important in an era of austerity. Cost-imposing strategies take actions

[6] Internet freedom can help the United States to compete against authoritarian rivals. Nevertheless, there is a tension between Internet freedom and other security concerns, particularly terrorism, which the revelations about the NSA surveillance program by Edward Snowden revealed.

[7] This figure includes only the State Department, USAID, and the BBG. Patricia Moloney Figliola, Kennon H. Nakamura, Casey L. Addis, and Thomas Lum, *U.S. Initiatives to Promote Global Internet Freedom: Issues, Policy, and Technology*, Washington, D.C.: Congressional Research Service, R41120, January 3, 2011, p. 16.

that pressure rivals to implement disproportionately costly countermeasures. Many current and potential authoritarian competitors, such as Iran and China, face endemic domestic tensions that center on a lack of freedom and regime legitimacy issues. As a result, both Tehran and Beijing have emphasized preserving domestic stability above all other goals. In short, promoting Internet freedom capitalizes on traditional American strengths and simultaneously exploits the enduring weaknesses of authoritarian rivals by compelling repressive regimes to spend ever-greater sums to preserve their rule. In an age of fiscal austerity and the growing economic prowess of potential adversaries, it is particularly important that the United States rely on policies, such as Internet freedom, that have a favorable cost-exchange ratio. At the same time, the United States must find ways to manage the challenge of allies that also try to censor and monitor the Internet.

Conclusions and Recommendations

In summary, we assessed DRL's Internet freedom portfolio to be balanced and of benefit to U.S. interests and values. Analyzing the portfolio's projects showed them to be properly targeted and executed and in line with DRL's overarching strategy. The overall contribution of the portfolio, if properly developed, should increase in the future. The portfolio appeared to be balanced, with a healthy mix of objectives and approaches. Risk seemed appropriate and prudent, with a suitable degree of failure tolerance spread across the portfolio and counterbalanced by a significant investment in lower-risk projects.

There are clear indications that Internet freedom has a positive, but indirect, connection to enlarging political space within repressive regimes.[8] Internet freedom initiatives also have the potential to be a high-leverage national security tool for democratic open societies, with cost-imposing characteristics against authoritarian regimes.

---

[8] This conclusion is drawn from a stand-alone study completed by RAND on the relationship between Internet freedom and political space. See Olesya Tkacheva, Lowell H. Schwartz, Martin C. Libicki, Julie E. Taylor, Jeffrey Martini, and Caroline Baxter, *Internet Freedom and Political Space*, Santa Monica, Calif.: RAND Corporation, RR-295-DOS, 2013.

As a result of our assessment, we developed four recommendations for DRL to consider as their portfolio continues to mature. Our first recommendation, which DRL began to address during the course of the assessment, is to enhance the synergy within the portfolio and among its grantees. As discussed, this is the area that would provide the largest return on investment for DRL. It would increase the effectiveness of the entire portfolio while decreasing its management and performance risk. Beyond merely providing opportunities for intraportfolio collaboration, we also recommended that DRL create mechanisms to incentivize collaboration and that these be outlined in the requests for proposals. Additionally, we encourage DRL and other USG agencies working on Internet freedom to explore ways to increase collaboration among the broader USG-funded Internet freedom community.

Second, we urge DRL to maintain a relatively balanced Internet freedom strategy that includes projects working on access, anonymity, awareness, and advocacy. While it may be necessary that one or several factors are given priority at a particular time, DRL should remain active in all four areas. It is increasingly apparent that circumvention and anonymity technologies alone cannot preserve Internet freedom. Instead, it is important both to train the at-risk individuals in how to use these technologies and to support efforts that uphold a free and open Internet in the domestic and international political arenas.

Third, we recommend that DRL consider a resourcing mechanism for contingency tasking. This recommendation grew from observing that several of the projects were actively seeking ways that they could leverage their DRL funds or their work for DRL to respond to the rapidly deteriorating situation in Syria. This demonstrated that a key value of the DRL portfolio was the residual capability of the network that it helped build. Having a way to rapidly and robustly energize that network to respond to other emerging crises would provide the government with a unique soft-power tool.

Fourth, we recommend that DRL consider this assessment as a rigorous first look at its portfolio, but to fully realize its value, this process should be repeated over time. This is a one-time assessment, which accurately represents a snapshot of DRL's Internet freedom portfolio in fiscal year 2012–2013. As we have discussed, circumstances and strategies change—often very rapidly in this arena. Therefore, the Internet freedom portfolio needs to be periodically reassessed to monitor its response to these changes and to ensure that it is still optimized to achieve the State Department's objective of expanded political space.

Acknowledgments

The authors thank Ian Schuler, Chris Riley, Betsy Bramon, and Stephen Schultze from the U.S. Department of State for their valuable feedback and guidance on this study. At RAND, we gratefully acknowledge the help of our colleagues Olesya Tkacheva and Christina Bartol. Special thanks to Eric Schwab for his assistance with part of the data analysis. Finally, we thank our reviewers, Richard Fontaine, Andrew Morral, and Sasha Romanosky, for their helpful comments.

Abbreviations

BBG	Broadcasting Board of Governors
DARPA	Defense Advanced Research Projects Agency
DRL	Bureau of Democracy, Human Rights, and Labor
ICANN	Internet Corporation for Assigned Names and Numbers
IP	Internet protocol
IT	information technology
NGO	nongovernmental organization
PortMan	Portfolio Analysis and Management Method
SME	subject matter expert
Tor	The Onion Router
USAID	United States Agency for International Development
USG	U.S. government
VPN	virtual private network

CHAPTER ONE

Introduction

Diverging Views on the Impact of the Internet

Currently, there is a struggle between those who want to communicate freely and securely and those who seek to restrict the content available online and to monitor and control the Internet. New communications technologies have converted the Internet from primarily a static vehicle for consuming information into an interactive cyber community.[1] The Internet is increasingly dominated by social media platforms and tools that enable users to produce online content, interact with others, and coordinate their actions. As social media sites—which include microblogging, social networking, photo and video sharing, social news, and virtual gaming websites—have proliferated, it has created new opportunities for collaboration and mass mobilization.[2] While most people use social media sites as a diversion, these online platforms can also strengthen civil society by spreading information and encouraging debate. Additionally, social media can be used as a coordinating tool,

---

[1] Sarah Joseph, "Social Media, Political Change, and Human Rights," *Boston College International and Comparative Law Review*, Vol. 35, No. 1, January 1, 2012, pp. 145–150.

[2] Microblogging sites include Twitter and Sina Weibo. Examples of social networking sites are Facebook and MySpace. Instagram is a hybrid photo and video sharing and social networking website. Other image sharing sites include YouTube and Flickr. Social news sites, such as Reddit and Digg, contain user-posted stories and allow users to rank the popularity of these stories and to comment on them. Virtual gaming websites include the World of Warcraft and EverQuest.

helping groups to overcome collective action challenges and ultimately to effect social and political change.[3]

The ability of social media to empower people was dramatically illustrated during the Arab Spring uprisings of 2011 and 2012.[4] In Tunisia, for example, the self-immolation of a young street vendor sparked demonstrations in the town of Sidi Bouzid. Tunisian protestors used cell phones to capture photos and videos of the brutal police response, then posted them on Facebook and YouTube. Because Tunisia had a relatively high rate of Internet penetration, these images rapidly spread across the nation, galvanizing citizens who were outraged by their government's brutality, corruption, and incompetence to stage additional protests.[5] Eventually, more than 2 million Tunisian members of Facebook changed their profile picture to read "Ben Ali dégage!" (Ben Ali Get Out!).[6]

---

[3] Clay Shirky, "The Political Power of Social Media: Technology, the Public Sphere, and Political Change," *Foreign Affairs*, Vol. 90, No. 1, January/February 2011, p. 29; Clay Shirky, *Here Comes Everybody: The Power of Organizing Without Organizations*, New York: Penguin Books, 2011; Eric Schmidt and Jared Cohen, "The Digital Disruption: Connectivity and the Diffusion of Power," *Foreign Affairs*, Vol. 80, No. 6, November/December 2010; Joseph, 2012, pp. 152–156; Emily Parker, *Now I Know Who My Comrades Are: Voices from the Internet Underground*, New York: Sarah Crichton Books, 2014.

[4] Philip N. Howard and Muzammil M. Hussain, "The Role of Digital Media," *Journal of Democracy*, Vol. 22, No. 3, 2011, pp. 35–36; Philip N. Howard and Muzammil M. Hussain, *Democracy's Fourth Wave: Digital Media and the Arab Spring*, Oxford, UK: Oxford University Press, 2013, p. 104. For a more skeptical view of the power of social media, see Malcolm Gladwell, "Small Change: Why the Revolutions Will Not Be Tweeted," *New Yorker*, October 4, 2010. For a more nuanced look at the effect of social media on the Arab Spring that focuses on causal mechanisms and empirical tests, see Sean Aday, Henry Farrell, Marc Lynch, John Sides, John Kelly, and Ethan Zuckerman, *Blogs and Bullets: New Media in Contentious Politics*, Peaceworks No. 65, Washington, D.C.: United States Institute of Peace Press, September 2010; Sean Aday, Henry Farrell, Marc Lynch, John Sides, and Deen Freelon, *Blogs and Bullets II: New Media and Conflict After the Arab Spring*, Peaceworks No. 80, Washington, D.C.: United States Institute of Peace Press, July 2012.

[5] Sanja Kelly and Sarah Cook, eds., *Freedom on the Net 2011: A Global Assessment of Internet and Digital Media*, Washington, D.C.: Freedom House, April 18, 2011, pp. 322–323.

[6] Marc Lynch, *The Arab Uprising: The Unfinished Revolutions of the New Middle East*, New York: PublicAffairs, 2013; Laryssa Chomiak and John P. Entelis, "The Making of North Africa's Intifada," in David McMurray and Amanda Ufheil-Somers, eds., *The Arab Revolts:*

At the same time, states can use the Internet to monitor and control their citizens. Some critics argue that the Internet and social media do not inevitably spread democracy (as the Arab Spring uprisings have demonstrated), but instead help to entrench authoritarian regimes.[7] According to this view, while the Internet may seem to empower activists, it disproportionately benefits governments that are better resourced and can exploit online information to monitor and control their citizens. In short, the Internet is more of a tool of repression than popular liberalization because it enables governments to cheaply and easily restrict their citizens' freedoms, locate and track down protesters, and spread their own propaganda. For example, in 2011, the Syrian regime removed its ban on Facebook and other popular social networking sites so that it could more easily monitor opponents' communications.[8] Relaxing censorship, therefore, enabled supporters of Bashar al-Assad to launch phishing attacks against dissidents and to infiltrate their networks.[9] Thus, the Internet is not inherently emancipatory; it can also increase oppression.

Recognizing that "modern information networks . . . can be harnessed for good or ill" and that "a new information curtain" had descended "across much of the world," then Secretary of State Hillary Clinton announced in January 2010 that "we stand for a single Internet where all of humanity has equal access to knowledge and ideas." Clinton also pledged that the United States would work to ensure that these technologies are "a force for real progress in the world."[10]

Dispatches on Militant Democracy in the Middle East, Bloomington, Ind.: Indiana University Press, 2013, p. 46.

[7] Evgeny Morozov, *The Net Delusion: The Dark Side of Internet Freedom*, New York: PublicAffairs, 2011; Rebecca MacKinnon, *Consent of the Networked: The Worldwide Struggle for Internet Freedom*, New York: Basic Books, 2012.

[8] Sanja Kelly, "Despite Pushback Internet Freedom Deteriorates," in Sanja Kelly et al., *Freedom on the Net 2013: A Global Assessment of Internet and Digital Media*, Washington, D.C.: Freedom House, 2013, p. 684; Joe Kloc, "Syria Grants Free Internet Access So It Can Snoop," *Newsweek*, February 27, 2014.

[9] Eva Galperin and Morgan Marquis-Boire, "Syrian Activists Targeted with Facebook Phishing Attack," *Electronic Frontier Foundation*, March 29, 2012.

[10] Elizabeth Dickinson, "Internet Freedom: The Prepared Text of U.S. Secretary of State Hillary Rodham Clinton's Speech, Delivered at the Newseum in Washington, D.C.," *Foreign Policy*, January 21, 2010.

The U.S. government (USG) has awarded Internet freedom grants since the early 2000s. Initially, most USG-sponsored efforts were focused on helping individuals to bypass firewalls in such societies as China and Iran.[11] Since 2008, the Department of State has broadened its program by funding Internet freedom as a part of its global human rights agenda with the goal of "ensur[ing] that any child, born anywhere in the world, has access to the global Internet as an open platform on which to innovate, learn, organize, and express herself free from undue interference or censorship."[12] The U.S. Department of State's Bureau of Democracy, Human Rights, and Labor (DRL) Internet freedom program awards grants to groups that are trying to advance Internet freedom by countering censorship, developing secure ways to communicate, providing digital safety training, conducting research on the effects of Internet freedom and Internet repression, and directly supporting activists on the front lines of the struggle against authoritarian regimes.[13]

Summary of Methodology and Findings

RAND researchers conducted an assessment of the DRL Internet freedom portfolio for fiscal year 2012–2013. Employing portfolio analy-

---

[11] Thomas Lum, Patricia Moloney Figliola, and Matthew C. Weed, *China, Internet Freedom, and U.S. Policy*, Washington, D.C.: Congressional Research Service, R42601, July 13, 2012, p. 12.

[12] U.S. Department of State, *Internet Freedom*, undated.

[13] U.S. Department of State, undated; Daniel Baer, Deputy Assistant Secretary, DRL, "Promises We Keep Online: Internet Freedom in the Organisation for Security and Cooperation in Europe (OSCE) Region," Statement Before the Commission on Security and Cooperation in Europe (U.S. Helsinki Commission), Washington, D.C., July 15, 2011. For additional information on the USG's efforts to promote Internet freedom, see Patricia Moloney Figliola, Kennon H. Nakamura, Casey L. Addis, and Thomas Lum, *U.S. Initiatives to Promote Global Internet Freedom: Issues, Policy, and Technology*, Washington, D.C.: Congressional Research Service, R41120, January 3, 2011; Patricia Moloney Figliola, *Promoting Global Internet Freedom: Policy and Technology*, Washington, D.C.: Congressional Research Service, R41837, April 23, 2013; and Richard Fontaine and Will Rogers, *Internet Freedom: A Foreign Policy Imperative in the Digital Age*, Washington, D.C., Center for New American Security, June 2011. Lum, Figliola, and Weed, 2012, pp. 12–13; Fergus Hanson, *Baked In and Wired: eDiplomacy@State*, Washington, D.C.: Brookings Institution, October 25, 2012, p. 26.

sis techniques, the assessment showed good alignment between DRL's strategy and the cumulative effect of the 18 funded projects. To assess DRL's Internet freedom portfolio, we began by developing an Internet freedom model that identified the variables that influence whether an individual chooses to use the Internet to expand political space. From this model, we created a set of metrics that formed the basis of the value, cost, and risk scores, which determined the portfolio's overall performance. To do this, RAND researchers built a survey protocol to ensure that its interviews of the DRL grantees were standardized and replicable. To assess DRL's Internet freedom portfolio, we collected information on each grantee through a semistructured interview and then relied on a small group of experts and an approach that facilitates consensus-building to estimate the value, risk, and cost metrics for each project.[14] The resulting data were then analyzed using the Portfolio Analysis and Management Method (PortMan) to determine the portfolio's overall performance, balance, and synergy. *Performance* indicates the portfolio's value, but also its risk and cost. For *balance*, we considered the type of projects that DRL funded, the geographic distribution of its projects, and the level of risk across the portfolio. *Synergy* refers to links between different grantees and explores whether the projects are working together to increase the performance of the overall portfolio.

In short, our assessment concluded that DRL's Internet freedom portfolio of projects is balanced among the following features:

- development of new technologies that enable individuals to circumvent Internet filters, protect an individual's identity, and enhance online security
- training programs that inform individuals about their vulnerabilities and teach them how to minimize their risk while online
- advocacy programs that delegitimize Internet censorship, promote a multistakeholder model of Internet governance, teach individuals to use the Internet to achieve political objectives, and assist activists *in extremis*

[14] Due to concerns about the security of the democracy and human rights activists who benefit from these programs, we agreed not to reveal any information about the specific projects.

- testing and evaluation projects that ensure that the circumvention tools that DRL is funding are of high quality and do not have critical vulnerabilities
- mixed projects that are involved in multiple areas.

Nevertheless, there is room for improvement, especially in terms of enhancing synergy within the portfolio. The Internet freedom community that DRL nurtures is potentially one of the most important and enduring outcomes because it is likely to have considerable value beyond the portfolio's funded lifespan. Therefore, DRL should continue to prioritize fostering connections between its grantees. Additionally, there is a positive, but indirect, connection between DRL's Internet freedom portfolio and the expansion of civic freedom within authoritarian regimes. Alone, access to the Internet is unlikely to produce popular revolutions that overthrow authoritarian regimes and result in the establishment of liberal democracies, but unfettered and secure Internet access is a critical enabler and accelerant that can help to achieve these objectives.[15] Finally, we determined that the value of such analysis is best realized over multiple stages of the portfolio's lifecycle. While this one-time assessment represents an accurate snapshot at a particular moment in time, understanding the full value of the DRL portfolio, and possible areas of concern, calls for periodic assessment over time to validate its findings. This is particularly important because Internet freedom is a relatively new programmatic area, as well as a rapidly changing environment.

The remainder of this report is divided into five chapters. Chapter Two explains the methodologies employed to assess DRL's Internet freedom portfolio. Chapter Three examines how DRL's Internet freedom portfolio performed on several different dimensions, and Chapter Four assesses the portfolio balance and synergy. Chapter Five incorporates some additional observations and lessons learned from the assessment. Chapter Six presents the findings, conclusions, recommendations.

---

[15] This conclusion is drawn from a stand-alone study completed by RAND on the relationship between Internet freedom and political space. See Olesya Tkacheva, Lowell H. Schwartz, Martin C. Libicki, Julie E. Taylor, Jeffrey Martini, and Caroline Baxter, *Internet Freedom and Political Space*, Santa Monica, Calif.: RAND Corporation, RR-295-DOS, 2013.

CHAPTER TWO

Methodology for Assessing the Portfolio

In conducting this assessment, we encountered several challenges. Most notably, Internet freedom is a relatively new area, which meant that we had to develop from the ground up a method for assessing DRL's portfolio. Moreover, due to the diversity of DRL's Internet freedom portfolio and the nature of the programs, there were no easily accessible or meaningful quantitative metrics for measuring the value of these projects. In part, this was because collecting project data could jeopardize the security of the activists who participate in these programs.

To overcome these challenges, RAND employed several methodologies, including a modified Delphi method and the PortMan framework. Figure 2.1 depicts the sequence in which these activities were carried out. First, RAND developed an Internet freedom model to identify the variables that affect an individual's willingness to use the Internet to expand political space *and* that can be influenced by the USG.

Second, to gather information on the 18 projects, we conducted interviews with each of the grantees and examined supporting materials, such as quarterly reports. The interviews covered a range of topics, including the project's background and objectives; activities to promote access, anonymity, awareness, and advocacy; programmatic execution; deployment strategy; staff credentials; and measures of performance. It was agreed with DRL that it would be beyond the scope of this project to measure the actual effect of these programs (e.g., test the technology, gauge what trainees actually learned), but we did consider the projects' execution and assessed whether they were hitting their cost, schedule, and performance targets (in terms of number of individuals

Figure 2.1
Methodology for Assessing DRL's Internet Freedom Portfolio

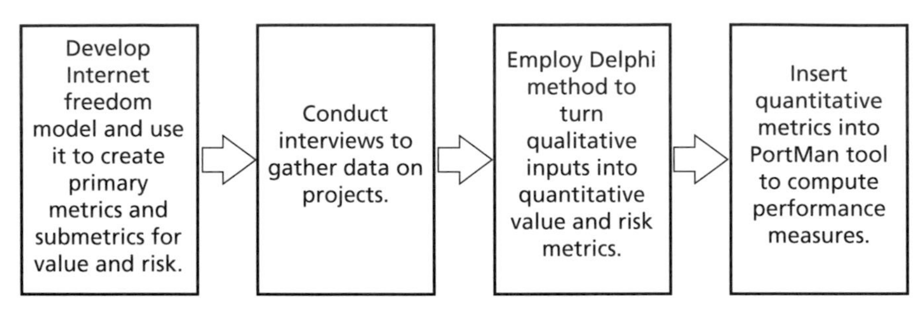

RAND *RR794-2.1*

trained, lines of code written, launching of a new website, and so on) that they had outlined in their statement of work. Third, the qualitative data gathered from these interviews was then converted into quantitative metrics through a modified Delphi method. Finally, these quantitative metrics were inserted into PortMan to determine the performance measure for each project. Each of these steps will be discussed in greater detail.

Developing the Internet Freedom Model

Our first task was to develop an Internet freedom model that captured the factors that influence whether a citizen in a repressive state decides to use the Internet to try to expand the political space in his or her society.[1] *Political space* consists of the freedom to assemble, free speech, and the ability to select a state's leaders through free and fair elections, and it expands when people exercise these rights. [2] In general, research has found that "the power of civil society is strengthened through higher levels of connectivity, unfettered access to knowledge, freedom

1 Initially, this model was developed through a deductive process, which identified the factors that influence whether a person uses the Internet to expand political space. The model was then refined through an iterative process with DRL, in particular, by adding the variable of advocacy.

2 Tkacheva et al., 2013, pp. 4–5.

of expression, and freedom to engage in collective action facilitated by digital tools: in short, the creation of social capital online."[3] Yet an individual decides whether to use the Internet for simply entertainment and personal communications or to achieve some sort of political objective. Our Internet freedom model is based on the assumption that individuals are essentially rational decisionmakers, meaning they consider the likelihood of realizing expected costs and benefits before acting and select the course of action that offers the highest expected payoff.[4]

When creating this model, we recognized that many of the factors that guide an individual's decision calculus are beyond the influence of the USG. For example, one of the critical contextual factors is the level of Internet penetration within a state.[5] A key precondition, therefore, is that a state has sufficient communications infrastructure in place so that citizens can go online. Additionally, there are other variables that may be affected by a combination of factors that are difficult to directly influence. Dissatisfaction with a repressive government, for example, is probably a product of each person's situation and personality.[6] Unhappiness with the regime is a necessary condition for an individual to use the Internet to try to increase openness in their state. Similarly, a person's level of risk acceptance and appetite for Internet freedom capabilities are likely to be strongly influenced by individual disposition. Some people are more willing to take dangerous actions, while others are

---

3 Robert Faris and Rebekah Heacock, "Introduction," in Urs Gasser, Robert Faris, Rebekah Heacock, eds., *Internet Monitor 2013: Reflections on the Digital World*, Cambridge, Mass.: Berkman Center for Internet and Society, December 12, 2013a, p. 1.

4 For more on rationality, see Jon Elster, ed., *Rational Choice*, New York: New York University Press, 1986. We realize that people are not perfectly rational and that misperception can often interfere with the pursuit of utility maximization. Nevertheless, it is very difficult to theorize about or predict misperception because it is often intimately linked to an individual's traits and experiences. For more on misperception, see Robert Jervis, *Perception and Misperception in International Politics*, Princeton, N.J.: Princeton University Press, 1976.

5 For various metrics on Internet penetration worldwide, see International Telecommunications Union, *The World in 2013: ICT Facts and Figures*, 2013.

6 Dissatisfaction or popular discontent is necessary but far from sufficient to explain activism or outright rebellion. For more, see Mark I. Linbach, *The Rebel's Dilemma*, Ann Arbor, Mich.: University of Michigan Press, 1998, p. 283. For more on discontent, see Ted Robert Gurr, *Why Men Rebel: Fortieth Anniversary Edition*, Boulder, Colo.: Paradigm Publishers, 2011.

more concerned about the potential negative repercussions and tend to avoid activity that could jeopardize their personal wellbeing.[7] An individual's predisposition to seek out Internet freedom technologies is probably tied to one's exposure to and general comfort with technology.[8] This, in turn, could also be related to age or many other individual traits, and therefore falls outside the realm of what the USG can directly influence. Finally, an individual's willingness to take the risky step of online activism will depend in part on that person's perception of whether the regime is capable of punishing him or her for this action and willing to do so.[9]

Although there are many variables that are outside the control of the USG, there are several critical factors—in particular, access, anonymity, awareness, and advocacy—that are amenable to manipulation (Figure 2.2). First, an individual who does not have access to the Internet is unlikely to be able to use it to expand political space. As mentioned above, Internet access requires infrastructure, but even if one can easily get online, many states still try to limit the content available by blocking or filtering websites considered to be unacceptable. For example, China uses automated filtering systems to restrict the websites that citizens can visit.[10] Other governments may try to inhibit access to information and online communications by shutting down or dramati-

---

[7] Varying risk acceptance has been shown to affect likelihood of collective action. Werner Raub and Chris Snijders, "Gains, Losses, and Cooperation in Social Dilemmas and Collective Action: The Effects of Risk Preferences," *Journal of Mathematical Sociology*, Vol. 22, No. 3, 1997, pp. 263–302.

[8] See similar research on why people use the Internet and why some choose not to go online: Kathryn Zickuhr, *Who's Not Online and Why*, Pew Research Center, September 25, 2013; Phys.org, "Factors Identified That Influence Willingness to Use Technology," March 8, 2013; and Edward D. Conrad, Michael D. Michalisin, and Steven J. Karau, "Measuring Pre-Adoptive Behaviors Toward Individual Willingness to Use IT Innovations," *Journal of Strategic Innovation and Sustainability*, Vol. 8, No. 1, June 2012, pp. 81–92.

[9] Mancur Olson, "The Logic of Collective Action in Soviet-Type Societies," *Journal of Soviet Nationalities*, Vol. 1, No. 2, Summer 1990, p. 16.

[10] Robert Faris and Nart Villeneuve, "Measuring Global Internet Filtering," in Ron Deibert, John Palfrey, Rafal Rohozinski, and Jonathan Zittrain, eds., *Access Denied: The Practice and Policy of Global Internet Filtering*, Cambridge, Mass.: MIT Press, 2008; and Jonathan Zittrain and John Palfey, "Internet Filtering: The Politics and Mechanisms of Control," in

Figure 2.2
Internet Freedom Model

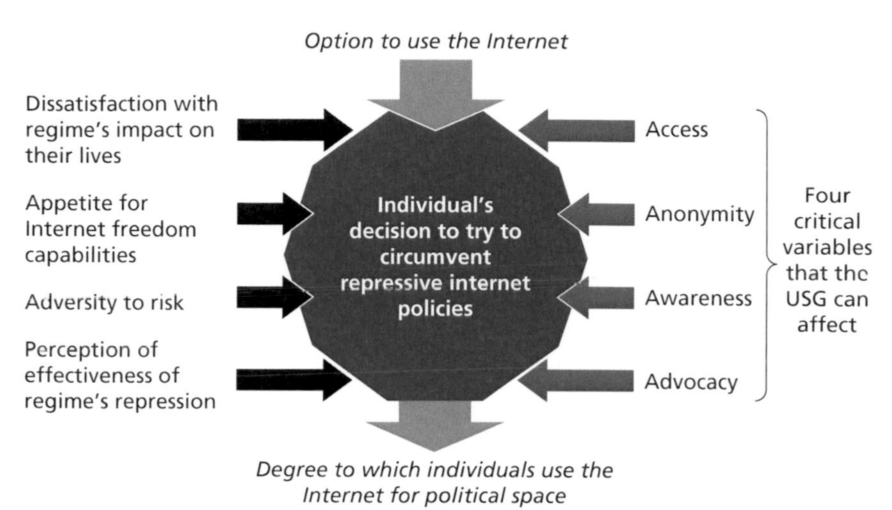

RAND *RR794-2.2*

cally throttling the speed of Internet connections.[11] Unfettered access to the Internet helps individuals to spread information about injustices within their state, to communicate with others, and to coordinate their actions.[12] In short, access is a key factor that influences whether one can use the Internet to expand political space.

Second, individuals are more likely to be politically active online if the government cannot easily identify and punish them. Yet government surveillance of the Internet is on the rise.[13] An ever-growing number of countries are monitoring online activity at Internet choke

Ron Deibert, John Palfrey, Rafal Rohozinski, and Jonathan Zittrain, eds., *Access Denied: The Practice and Policy of Global Internet Filtering*, Cambridge, Mass.: MIT Press, 2008, p. 32.

[11] Collin Anderson, "Dimming the Internet: Detecting Throttling as a Mechanism of Censorship in Iran," June 18, 2013.

[12] Shirky, 2011, p. 31.

[13] Kelly, 2013, pp. 7–9; Sanja Kelly, Sarah Cook, and Mai Truong, eds., *Freedom on the Net 2012: Global Assessment of Internet and Digital Media*, Washington, D.C.: Freedom House, 2012, p. 10; Reporters Without Borders, *Enemies of the Internet 2013 Report: Special Edition Surveillance*, Paris, France, 2013.

points, such as Internet exchanges and Internet service providers.[14] This is particularly important for nondemocratic states that rely on coercion to stay in power and monitor a variety of online platforms, including mobile phone calls, text messages, email, browsing histories, voice over internet protocol (IP) calls, and instant messages to control political opposition. Consequently, it is critical to protect the anonymity of online activists.

Third, digital activists are unlikely to be effective if they are not aware of their vulnerabilities to online surveillance, or if they lack knowledge about basic digital safety practices. Technologies that provide secure access to the Internet are essential, but they are useless if their intended users to do not know how to properly employ these circumvention and anonymity tools. Therefore, an important way to facilitate online activism is by expanding awareness of how a state can monitor online activities and sharing principles for reducing one's exposure to surveillance, particularly by employing anonymity tools.

Finally, the USG can support advocacy programs to teach people how to exercise their basic rights, to delegitimize online censorship, to assist activists who are being prosecuted by a repressive state, and to campaign for a multistakeholder model of Internet governance. Trends suggest that the battle for Internet freedom is increasingly being waged in legislatures, courts, and international institutions.[15] Passing laws that criminalize online speech or hold intermediaries liable for the content posted on their websites is a more insidious way of stifling online freedom because it encourages self-censorship. Beijing, for example, has created a climate of fear by requiring real-name registration (through identification cards with embedded computer chips) at Internet cafes and for a number of popular websites, such as the microblogging platform Sina Weibo.[16] Increasingly, governments are also arresting people and

---

[14] Ronald Diebert and Rafal Rohozinski, "Beyond Denial: Introducing Next Generation Information Access Controls," in Ronald Deibert, John Palfrey, Rafal Rohozinski, and Jonathan Zittrain, eds, *Access Controlled: The Shaping Power, Rights, and Rule in Cyberspace*, Cambridge, Mass.: MIT Press, 2010, p. 9.

[15] Faris and Heacock, 2013a, p. 8; Kelly, 2013, pp. 9–12.

[16] Open Net Initiative, *Profiles: China*, August 9, 2012.

handing out extremely harsh prison sentences for content that they have posted online.[17]

Similarly, there is a growing movement to shift from the current multistakeholder model of Internet governance to one that is controlled by nation-states. A growing number of states—led by China and Russia—are demanding that control over the Internet be transferred to a subcomponent of the United Nations, the International Telecommunications Union, which they can influence.[18] To counter these moves, the United States can back campaigns that support sensible domestic and international regulations for the Internet and that preserve the multistakeholder model.[19]

In sum, there are four critical variables—access, anonymity, awareness, and advocacy—that influence an individual's propensity to use the Internet to expand political space and that are also amenable to intervention by the USG. Consequently, in addition to cost and risk, these four variables are the primary metrics that we used to assess the value of DRL's Internet freedom program. The methodologies used to evaluate the program are discussed in greater detail in the next section.

---

[17] Kelly, 2013, p. 10.

[18] Robert Faris and Rebekah Heacock, "Looking Ahead," in Urs Gasser, Robert Faris, and Rebekah Heacock, *Internet Monitor 2013: Reflections on the Digital World*, Cambridge, Mass.: Berkman Center for Internet and Society, December 12, 2013b, pp. 86–87. For more on Internet governance, see Laura DeNardis, *The War for Internet Governance*, New Haven, Conn.: Yale University Press, 2014.

[19] As a part of this effort, in March 2014, the U.S. Department of Commerce announced that it was allowing its contract with the Internet Corporation for Assigned Names and Numbers (ICANN) to lapse and transitioning oversight of the web addresses and IP numbers to a multistakeholder model. This move was intended to defuse criticism that the United States maintains too much control over the Internet's architecture and to undermine support for the campaign to transfer power to the International Telecommunications Union. For more, see Stacie L. Pettyjohn, "Net Gain: Washington Cedes Control Over ICANN," *Foreign Affairs*, April 10, 2014.

Developing Nested Measures of Value, Risk, and Cost to Input into the PortMan Analysis

The core of our analysis was built around the RAND Corporation's PortMan framework to evaluate DRL's Internet freedom program portfolio.[20] Portfolio analysis involves assessing "the contributions and balance of a collection of projects aimed at achieving a common goal" and differs from an independent evaluation of each project against its own objectives.[21] RAND's PortMan allows one to monitor performance and to make data-driven decisions, which, if used over time, can assist in realizing the highest expected value from a portfolio.[22] PortMan also helps to ensure that a portfolio is appropriately balanced and that it aligns with the organization's overall objective.

To use PortMan, we must estimate the value and risk of each project based on an agreed-upon set of metrics. The expected value of a project is a product of the expected value if a project is successfully implemented and the risk or probability of successful implementation.[23] We used the previously discussed Internet freedom model to identify metrics that were used to estimate the value of each of the projects. The Internet freedom performance value of the individual projects was determined by each project's overall contributions to four major variables that affect Internet freedom and political space: (1) access to the Internet, (2) anonymity and security when accessing the Internet, (3) awareness and understanding of security threats and protective measures, and (4) advocacy for ensuring a free and open Internet com-

---

[20] Alan A. Lewis, *The Use of Utility in Multiattribute Utility Analysis*, Santa Monica, Calif.: RAND Corporation, P-6396, 1980. For more on PortMan, see Richard Silberglitt, Richard, Lance Sherry, Carolyn Wong, Michael Tseng, Emile Ettedgui, Aaron Watts, and Geoffrey Stothard, *Portfolio Analysis and Management for Naval Research and Development*, Santa Monica, Calif.: RAND Corporation, MG-271-NAVY, 2004; Eric Landree, Richard Silberglitt, Brian G. Chow, Lance Sherry, and Michael S. Tseng, *A Delicate Balance: Portfolio Analysis and Management for Intelligence Information Dissemination Programs*, Santa Monica, Calif.: RAND Corporation, MG-939-NSA, 2009.

[21] Landree et al., 2009, p. 1.

[22] Landree et al., 2009, p. 5.

[23] Landree et al., 2009, pp. 5–6.

patible with the aforementioned U.S. policy toward free speech and human rights online.

A project's risk level and cost, both direct and indirect, to the USG also influenced its portfolio contribution. Risk is defined as the probability of successful implementation and indicates the difficulty of executing and sustaining a project.[24] Risk was assessed in three areas: (1) capability, or the reliability of a project's planned approach;[25] (2) acceptance, which assessed a project's credibility with users and deployment plan for the product; and (3) sustainability, which considered the project's planning for future activities and funding. Cost elements included a project's level of funding, as well as potential indirect costs through management burden or international political exposure.

The primary metrics of value, risk, and cost were each divided into subcomponents developed in consultation with the RAND team, the State Department, and area experts that could be measured and inserted into PortMan. As will be discussed in greater detail below, each subcomponent was scored by a small group of experts on a scale from one to five. In total, experts scored each project on 29 separate subcomponents, which were then aggregated to determine a project's overall performance score.[26]

As Figure 2.3 depicts, total performance is composed of risk, value, and cost. Value, in turn, is made up of access, anonymity, awareness, and advocacy, and each of these four primary elements of value consists of several subcomponents. Access, for example, includes the

---

[24] Landree et al., 2009, p. 6.

[25] These projects were in various stages of execution; two were just beginning, and 16 were in the middle of executing their tasks when they were interviewed. To deal with this fact, the capability risk measure differentiated between proven capability risk (or achievements thus far) and prognostic capability risk (or solidity of approach, staffing, etc.). The total capability risk score, therefore, was calculated based on how far along the project was. For instance, if a project was complete, its capability risk was composed entirely of proven measures, while if it was only partially executed, its score would be half proven, half prognostic.

[26] The aggregate score was calculated by taking the log of the relevant subcomponents and then normalizing the score. So, for an element E with three subcomponent measures A, B, and C, the element score was $E = (\log^5 (A^*B^*C))/3$. We employed this methodology to accommodate the one-to-five normalized scoring range for all principal measures and enable us to aggregate submeasures independent of their ordering in a consistent manner.

Figure 2.3
Nested Measures Constitute Total Performance

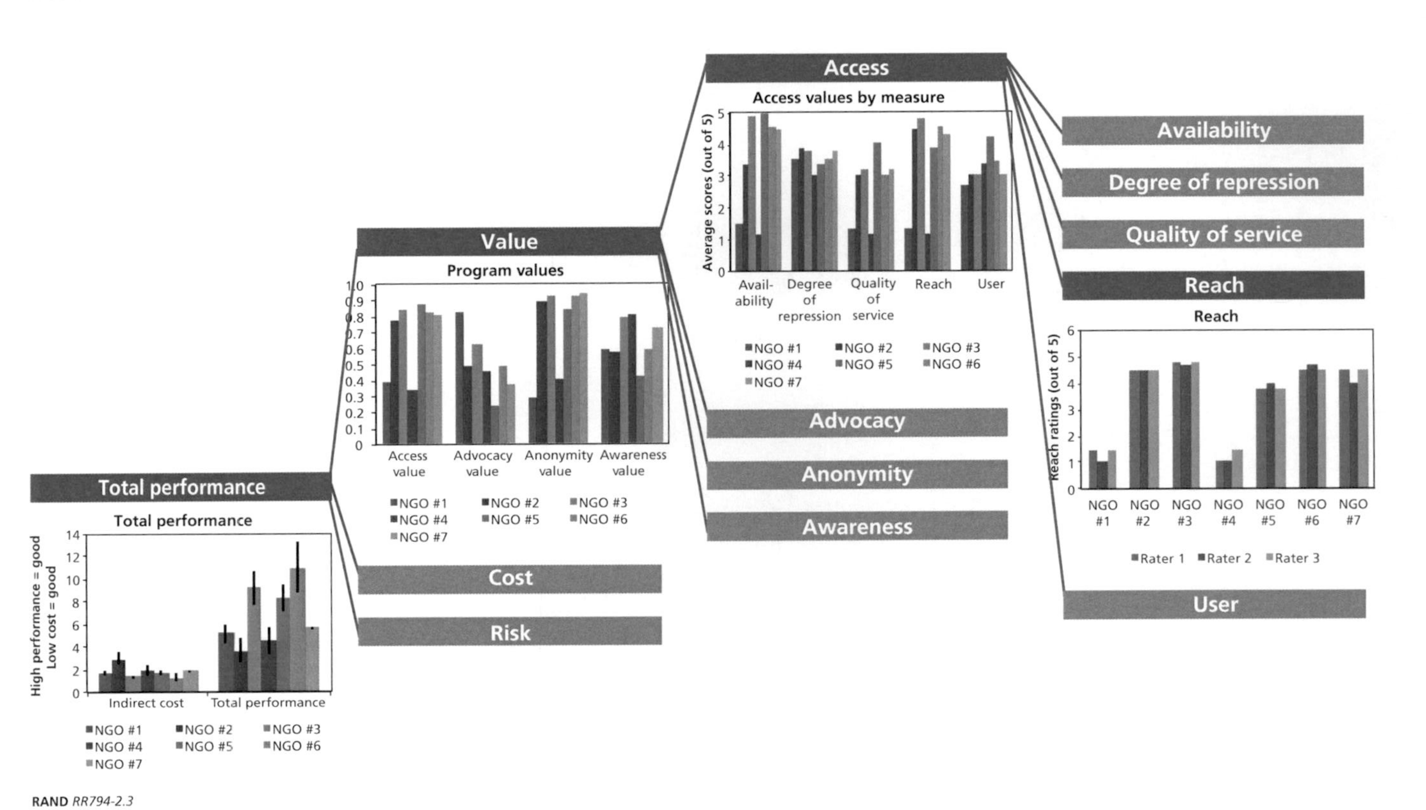

RAND *RR794-2.3*

user skill level, degree of repression in target state, reach, quality of service, and availability. The remaining components of value and their subcomponents are discussed in detail in Chapter Three. Total performance, therefore, is calculated by measuring a series of components and subcomponents that are nested within each other.

Gathering Data and Transforming Qualitative Inputs into Quantitative Measures

To gather data on DRL's Internet freedom projects, RAND researchers conducted semistructured interviews with each project team. Whenever practicable, the interviews were in person. To ensure that the interviews were standardized and carried out in a replicable manner, we developed an extensive survey protocol, which we used in each of the interviews. The protocol included questions that would elicit information that we could use to estimate the expected value and risk for each project, including its background, desired outcomes, specific outputs, implementation strategy, methodology, alignment with key Internet freedom attributes, cross-project synergy, tool employment, measures of performance achievement, and measures of effectiveness relevance and technical, programmatic, and acceptance risk.

We relied on a small group of subject matter experts (SMEs) and an approach that facilitates consensus-building to estimate the value and risk metrics for each project.[27] The SMEs assessed each of the DRL Internet freedom projects on 29 dimensions.[28] We used some of our own experts in a modified Delphi consensus-building exercise to estimate the value and risk scores for each project.[29] To employ the

---

[27] For this project, four RAND staffers who had expertise on Internet freedom and DRL's program served as SMEs. Silberglitt et al., 2004; Landree et al., 2009; Richard Silberglitt and Lance Sherry, *A Decision Framework for Prioritizing Industrial Materials and Research and Development*, Santa Monica, Calif.: RAND Corporation, MR-1558-NREL, 2002.

[28] The value score consisted of 17 different dimensions, the risk score had 10 dimensions, and the cost score had two dimensions.

[29] Olaf Helmer-Hirschberg, *Systematic Use of Expert Opinions*, Santa Monica, Calif.: RAND Corporation, P-3721, 1967.

Delphi method, we provided each SME with a *dossier*—a collection of standardized materials—for each Internet freedom project, which was created from the data gathered during the interviews. In addition, we provided SMEs with a scoring guide outlining and defining each of the metrics and its range of scores. The Delphi consensus exercise typically consisted of three rounds, depending on how much variation existed among the SMEs' scores. Each SME scored a project separately, and then all of the SMEs gathered to discuss their assigned scores. In these discussions, particular attention was given to the rationale behind each score, as well as minority views. After the discussions, SMEs could individually change their scores based on what they had heard in the discussions (see Figure 2.4).[30] Consensus was not necessarily reached on every metric, but this was reflected in the uncertainty assigned to each score. As a final check on outliers, we shared these scores with the DRL grant officer representative assigned to the project to corroborate the findings from the Delphi method. In no instances did this result in modifications to the final scores.

Figure 2.4
Modified Delphi Method

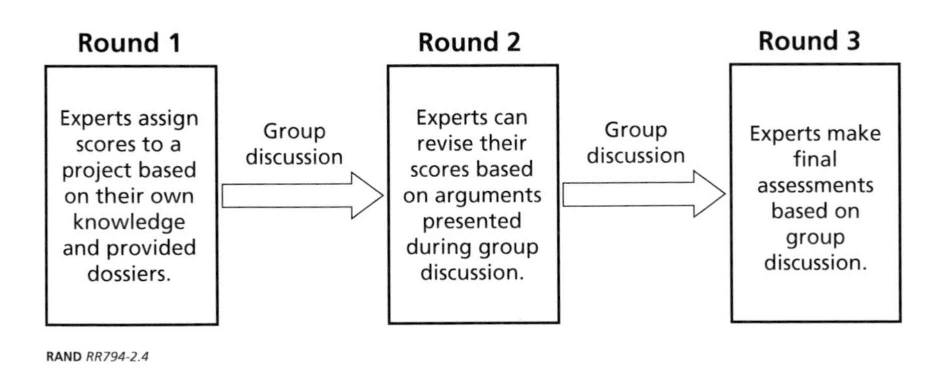

RAND *RR794-2.4*

[30] This departed from the traditional Delphi method in that participants (in this case the SMEs) were not anonymous. The traditional Delphi method, however, does have drawbacks, which include the lack of live discussions and the fact that it is time consuming. For these reasons, we used a modified Delphi method where participants knew each other's identities and participated in discussions, although they still independently made their assessments (i.e., scores). Silberglitt et al., 2004, p. 23. For a traditional Delphi method, see Norman Crolee Dalkey, *The Delphi Method: An Experimental Study of Group Behavior*, Santa Monica, Calif.: RAND Corporation, RM-5888-PR, 1969.

Using PortMan to Calculate Performance

This Internet freedom portfolio assessment was based on the cumulative performance value of individual projects within the portfolio. The major component scores (normalized to be out of one) in each area were then combined in accordance with PortMan to produce a single performance measure for each program. Under the PortMan framework, performance is a function of value added and programmatic measures—in this case, risk and cost. The elements were combined according to the following formula (note that weighting of measures depends on DRL's strategy):[31]

$$\begin{aligned} \text{Performance} &= \text{portfolio contribution} \\ &= f(\textit{Value Measures}, \textit{Programmatic Measures}) \\ &= \frac{(\textit{Access} + (2 \times \textit{Anonymity}) + \textit{Awareness} + (\textit{Advocacy} \div 2)) \times (\textit{Risk})}{\textit{USGCost}} \end{aligned}$$

It is important to note that the four value measures (access, anonymity, awareness, and advocacy) were not given equal weighting in the formula. Instead, they were weighted in accordance with the DRL program strategy, which emphasized anonymity when using the Internet as a key component, and viewed advocacy as a less essential aspect.

During fiscal year 2012–2013, DRL's strategy to enhance Internet freedom consisted of four major components:

- countering online content restrictions by developing anticensorship technologies and expanding access to information and communications platforms
- developing secure communications technologies to strengthen privacy and anonymity online
- teaching individuals about good digital safety practices through training

---

[31] This approach was employed instead of such alternatives as net present value because DRL's Internet freedom program was not seeking to get a return on investment in terms of profit. Rather, the benefits yielded by these programs are largely intangible.

- advocating for national and international policies that protect Internet freedom.

Although all of these elements are important, DRL noted in its 2012 solicitation for proposals that "the greater use of Internet monitoring" and the fact that "secure communications have themselves become targets of conscious blocking efforts," there is a growing need for technologies and tools that enable "secure, private, or anonymous communications."[32] With these considerations, DRL's strategy prioritized efforts to enhance anonymity and enable secure communication over the other aspects of Internet freedom. Moreover, while access and awareness were equally valued, advocacy was considered to be less critical and, therefore, did not contribute as much to a project's overall performance.[33]

The resulting data were further analyzed using PortMan to determine the overall portfolio's performance and effectiveness. The PortMan analysis plots measures of value against measures of risk and cost. In Figure 2.5, a hypothetical program distribution is shown, with value measures increasing on the y-axis and programmatic measures of risk and cost decreasing along the x-axis. In this visualization, performance scores are largest in the top-right quadrant of the graph. Additionally, curved lines show the contours along which performance is equivalent (a project that is high-value and high-risk may have the same performance score as a project that is low-value and low-risk). The size of the circles depicts the uncertainty about the score, with larger circles indicating greater uncertainty. Uncertainty reflects the amount of disagreement among the SMEs over the measures.

Unless otherwise indicated, the data shown in this report reflect performance scores based on the existing DRL strategy. Under this methodology, the highest-performing projects were active in multiple areas of Internet freedom at comparatively low risk and cost. Addi-

---

[32] Grantsolutions.gov, "Bureau of Democracy, Human Rights and Labor and Bureau of Near Eastern Affairs Joint Request for Statements of Interest: Internet Freedom Programs," May 2012.

[33] The DRL program strategy was communicated to the RAND team in a series of meetings in which DRL proposed and approved the value weighting shown in Figure 2.5.

Figure 2.5
Stylized PortMan Analysis

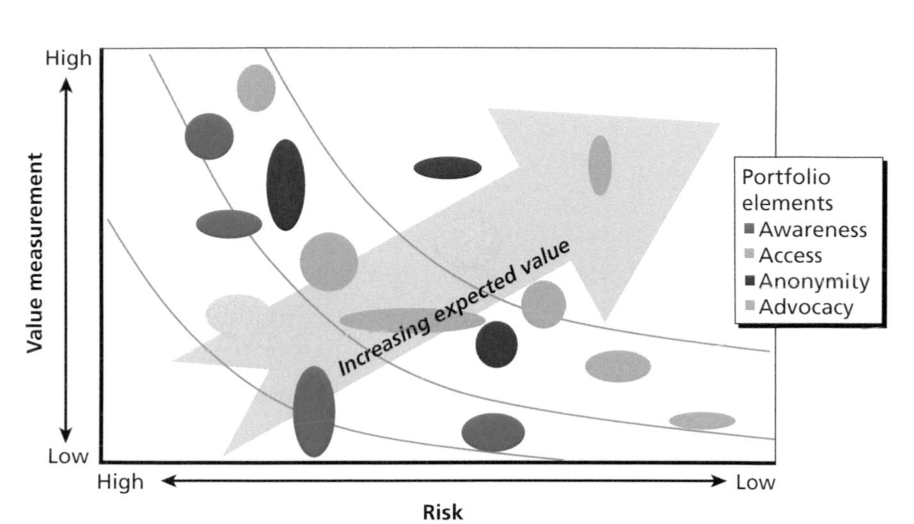

RAND *RR794-2.5*

tionally, because of the de-emphasis on advocacy, projects that focused primarily on advocacy efforts without incorporating other elements of Internet freedom received relatively low performance scores. Conversely, projects that made a key contribution to anonymity on the Internet were among the higher performers.

We used DRL's stated strategy to determine the performance formula because a key objective of our analysis was to determine if DRL was implementing its desired strategy. This decision came with some drawbacks. In particular, it assumes that DRL has developed a sound and effective strategy and, therefore, does not question the basic assumptions behind DRL's approach. For instance, the USG's Internet freedom program has been criticized for overemphasizing circumvention technologies while neglecting concerns about privacy and training programs that teach activists how to safely employ these technologies.[34] Our analysis does not get at fundamental questions about

---

[34] Fontaine and Rogers, 2011, pp. 37–38; Ethan Zuckerman, "Internet Freedom: Beyond Circumvention," *My Heart's In Accra* (blog), February 22, 2010; Leslie Harris et al., "An Open Letter to Congress About Internet Freedom," web page, March 14, 2011.

DRL's underlying strategy. Yet this approach does have the virtue of forcing DRL to explicitly discuss and identify its strategy and allows one to evaluate whether the declared strategy is being put into practice in DRL's Internet freedom portfolio. For illustrative purposes, we have also calculated the portfolio's performance using different strategies and performance formulas and shared these with DRL.

Moreover, because the competition between those trying to promote Internet freedom and those trying to control and monitor the Internet is rapidly evolving, DRL will likely have to frequently adapt its strategy to changing circumstances. A useful feature of PortMan is that, if the program strategy is altered, this weighting can be modified to generate new scores that measure project performance against the revised strategy.

CHAPTER THREE

Portfolio Performance

RAND researchers based the Internet freedom portfolio assessment on the cumulative performance value of individual projects within the portfolio. The performance value was determined by each project's overall contributions to the four major variables that affect Internet freedom and political space: (1) access to the Internet, (2) anonymity and security when online, (3) awareness and understanding of security threats and protective measures, and (4) advocacy for ensuring a free and open Internet.

Access

The access variable assessed a project's contribution to a user's ability to enjoy unfettered access to the Internet. Of the four value categories, this factor was the most complex, incorporating the following five subcomponents:

- user skill level—a measure of how technologically knowledgeable a user would need to be to maximize the project benefit
- degree of repression—a static measure of the environment of Internet repression by the local regime where a product would be deployed[1]
- reach—a measure of the degree of Internet access offered by a product in enabling a user to overcome censored or blocked sites

[1] The repression scores for countries and regions were derived from Kelly, 2013.

- availability—the percentage of time a product was available and functioning as a circumvention tool, particularly with respect to whether there were any outages in service or if it would function in the event of a full Internet shutdown
- quality of service—a measure of the usability of a product under the intended conditions or environment where it would be deployed (in particular, whether use of the product would affect latency, error rate, etc.).

In general, projects that performed well in the access category were involved in either development or distribution of circumvention technology. Of the 18 projects in the DRL portfolio, six were classified as development projects, meaning their primary activity was producing a new or improved circumvention tool that would allow their users to bypass firewalls and access blocked websites. The most common method for accessing censored content is through a proxy, which sends requests to visit blocked webpages through an unblocked computer, thereby enabling the user in the censored country to bypass content restrictions. Proxies can vary significantly in their complexity—ranging from simple one-hop, web-based proxies to a network of proxies—and in the type of security that they provide to their users. Another popular circumvention tool is a virtual private network (VPN), which creates an encrypted tunnel between two computers that all online traffic moves through.

Six additional projects were engaged in distributing circumvention tools and training users how to operate them. Finally, several projects aimed to prevent censorship by providing vulnerable websites with protection against distributed denial of service attacks, thereby ensuring that these sites were available for anyone to visit. Figure 3.1 depicts the access scores for the projects in DRL's Internet freedom portfolio.

An important aspect to note is that the access score was somewhat sensitive to the degree of online repression in a project's intended deployment locale. Projects that targeted especially repressive countries or regions (e.g., China or Iran) received a boost in their access score, reflecting the value added of providing circumvention technology in areas where it was most needed.

Figure 3.1
Project Access Scores

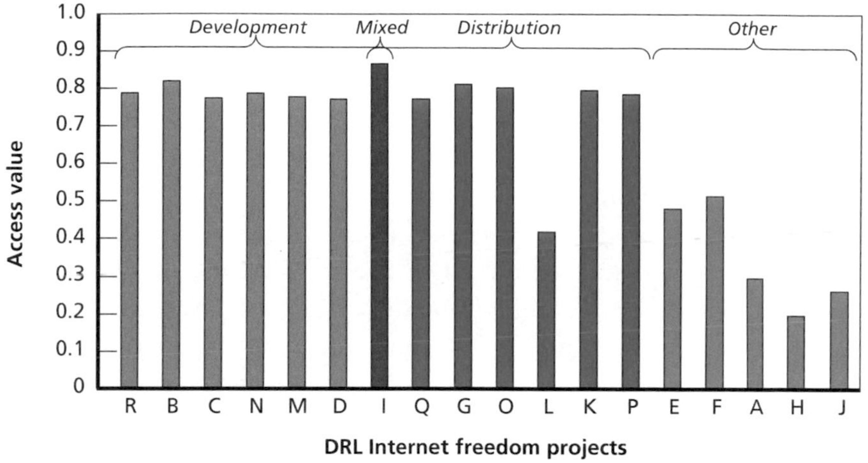

NOTE: Lettered project codes are used to maintain project anonymity.
RAND *RR794-3.1*

Anonymity

The anonymity variable assessed a project's contribution to a user's ability to securely and anonymously access Internet sites and send messages without regime visibility into the communications. Per the expressed DRL strategy, this factor was deemed the most important element of Internet freedom, as demonstrated by its double-weighting in the PortMan formula. The anonymity variable was divided into three subcomponents:

- visibility—a measure of the regime's ability to accurately detect and observe an Internet user's online activities when using the tools or techniques offered by the project
- attribution—a measure of the regime's ability to connect online activity with an Internet user's real identity when using the project offering
- localization—a measure of the regime's ability to accurately identify an Internet user's geographical location.

The majority of the projects active in the access category (producing or distributing circumvention technology) made similar contributions to anonymity, reflecting the growing recognition that unfettered access without security could be extremely dangerous for users. Many proxies and VPNs, therefore, encrypt traffic between the user and their server, which conceals the content of the messages and activity (i.e., reduces visibility) by scrambling the information so that it is readable only by those who have the key. It does not provide strong protection against attribution or localization, however, because the operators of these services can still identify their users. Moreover, interested third parties can still detect that an individual is using a circumvention tool.

By contrast, Tor (an anonymity technology based on proxy routing, also known as The Onion Router) provides strong safeguards against visibility, localization, and attribution. It does so by randomly sending Internet communications through a distributed network of proxies or relays. Tor not only encrypts all communications between the relays, but it also limits the amount of information that each relay has about the process. By compartmentalizing information about the Internet traffic, no one is able to conduct longitudinal traffic analysis or identify the user or her location. Nevertheless, there are some downsides to using Tor; it degrades the speed of the Internet connection and is somewhat complicated to install and employ. Because some users may be disinclined to use Tor for these reasons, it is important that the USG continue to develop multiple technologies to protect privacy.[2]

For our performance formula, the increasing use of mobile technology as the primary mode of Internet access worldwide posed particular challenges in the anonymity category (see Figure 3.2). Projects providing user trainings and distributing tools acknowledged the problem that while mobile technology is becoming more popular internationally, it is exceptionally vulnerable to regime monitoring with respect to localization and attribution. At present, there are no tools that provide a high level of anonymity when using mobile technology. But there are efforts to develop programs that provide greater privacy by encrypting

[2] Additionally, funding only one secure communication tool would ease the task of states trying to counter such technologies.

Figure 3.2
Project Anonymity Scores

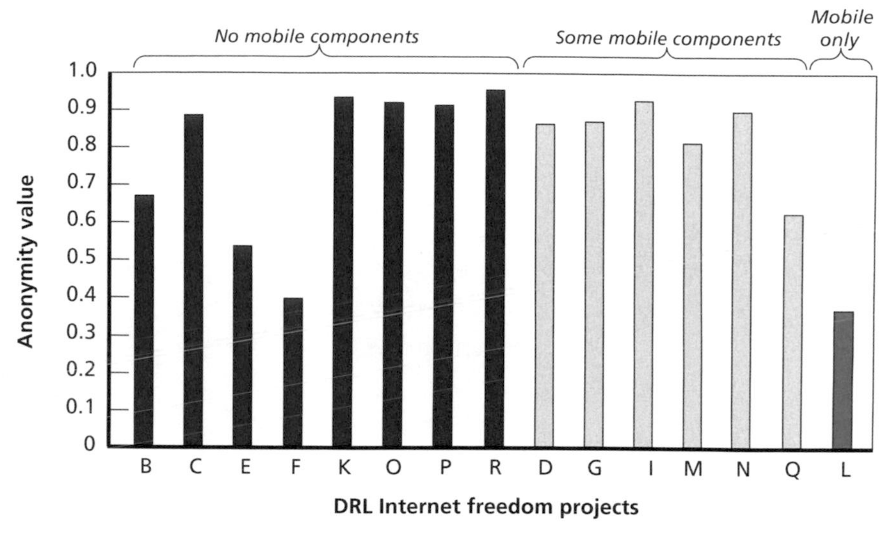

NOTE: Lettered project codes are used to maintain project anonymity.
RAND *RR794-3.2*

voice and text message communications over mobile phones. Other programs may try to conceal the fact that someone is sending a sensitive message by hiding text messages in seemingly benign content, like photographs, which is called *steganography*. If someone sends encrypted or hidden messages, a third party could still track the location of the phone and identify its user, but the content of the messages is not visible. Interestingly, the single mobile-focused project in the DRL programming opted not to distribute mobile circumvention technology to avoid giving users a false sense of security.

Awareness

The awareness variable was a measure of a project's contribution to a user's understanding of the sophistication of regime visibility into Internet use, as well as the measures that the individual user could take to increase his or her security. The awareness variable had the following three subcomponents:

- degree of monitoring—a static measure of the environment of Internet monitoring by the regime where a product would be deployed[3]
- user security awareness—a measure of the project's effect on the intended user's understanding of his or her vulnerability to regime monitoring.
- user circumvention awareness—a measure of the project's effect on the intended user's knowledge of the circumvention technology available and its appropriate usage.

Scores for user security and circumvention awareness were tabulated with a slightly different procedure. Because the emphasis for these measures was on project *effect* on user awareness, the measures were estimated for users before and after the project was implemented. The project score was then measured by the difference, or increase, in knowledge pre- and post-project.

As with access, including the static degree of monitoring measure meant that the geographic targeting of projects had an influence on their scores in this category. Projects that directed their efforts at countries with more repressive regimes received a scoring benefit that reflected the increased importance of user security awareness in these areas. Additionally, because of the way the influence criteria were scored here, projects whose intended users had a low understanding of regime monitoring prior to their interaction with the project had more potential for influence than projects that targeted highly knowledgeable users.

The highest performing projects in this area tended to be focused on training, or they incorporated training or public awareness campaigns as a major element of their project activities (see Figure 3.3). Programs focusing on security awareness tended to inform their trainees about their vulnerabilities when they are online or using their mobile phone and offered tips on how to reduce their risk. At the most

---

[3] The degree of monitoring scores for countries and regions were derived from Kelly, 2013.

Figure 3.3
Project Awareness Scores

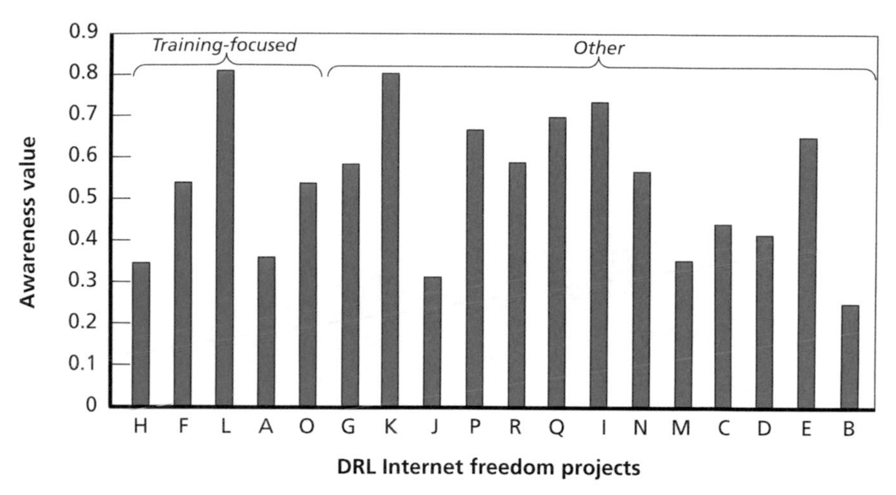

NOTE: Lettered project codes are used to maintain project anonymity.
RAND *RR794-3.3*

basic level, some programs emphasized the importance of using antivirus software and visiting secure (https) websites or using Google's email instead of Yahoo's email. At the other end of the spectrum, some programs educated activists on how to use multiple relatively complicated circumvention and anonymity technologies or practices simultaneously for added protection. Additionally, several programs aimed to identify their users' needs and to match them with circumvention tools that provided an appropriate level of protection given the risk associated with their online activities. For instance, some users may simply want to access blocked websites, such as Facebook or YouTube, for entertainment purposes. Because these users are not engaged in high-risk behavior, a proxy or VPN that provides some but not complete anonymity is probably sufficient. By contrast, citizen-journalists or activists who are organizing opposition to a repressive regime need tools that both allow them to access blocked content and protect their identity. In short, some programs found that, because it is often difficult to convince people to adopt practices that improve their online security, it is important to tailor the advice given and tools suggested to increase the likelihood that they are adopted.

We found that there was a high degree of diversity in the strategies that projects employed to train users, ranging from short, mobile microsessions in Internet safety to detailed, month-long courses. In addition, some project trainings were conducted in person, which often meant that individuals had to leave the country to attend trainings in a less repressive region; other projects opted to place all training materials online and teach courses in a virtual classroom. Both strategies have merits and drawbacks but may be tailored to the unique challenges faced in each region.

Advocacy

Advocacy was the final variable included in measuring project value, and it referred generally to project efforts that influenced Internet environments or that promoted the concept of a free and open Internet compatible with U.S. policy toward free speech and human rights online. The four subcomponents contributing to the advocacy measure were less interdependent than other categories; in other words, a program did not need to perform well in all four categories to contribute value. The four subcomponents, developed in discussion with DRL and based on their objectives, included the following:

- censorship and surveillance delegitimization—a measure of efforts to make political censorship and pervasive online monitoring unacceptable activities
- multistakeholder net governance—a measure of efforts to make net governance more inclusive of civil society participants
- assistance to activists *in extremis*—a measure of the provision of emergency or legal services to activists persecuted or prosecuted by a regime for online activity
- employment of Internet's political space—a measure of assistance and training of activists on effective techniques for using the Internet to expand political space.

We observed that projects with a focus in one or more of the other value categories made only small, second-order contributions to advocacy (see Figure 3.4). The four advocacy-specific projects in the DRL portfolio tended to be narrowly scoped and made fewer contributions in the other value categories, suggesting that there is less interaction between this element and the other three categories, at least in practice. Several of the projects in this category focused on conducting new research on the Internet and, particularly, on cataloging Internet controls in an effort to delegitimize them. Other advocacy-focused projects aimed to influence domestic legislation in countries that were at risk of implementing laws that would restrict online freedoms. Finally, some advocacy projects focused on influencing international standards by campaigning on behalf of a multistakeholder model of government. Because of the reduced weighting for advocacy in DRL's PortMan strategy, these projects tended to receive lower-than-average total performance scores, although the projects themselves seemed topical and well run.

Figure 3.4
Project Advocacy Scores

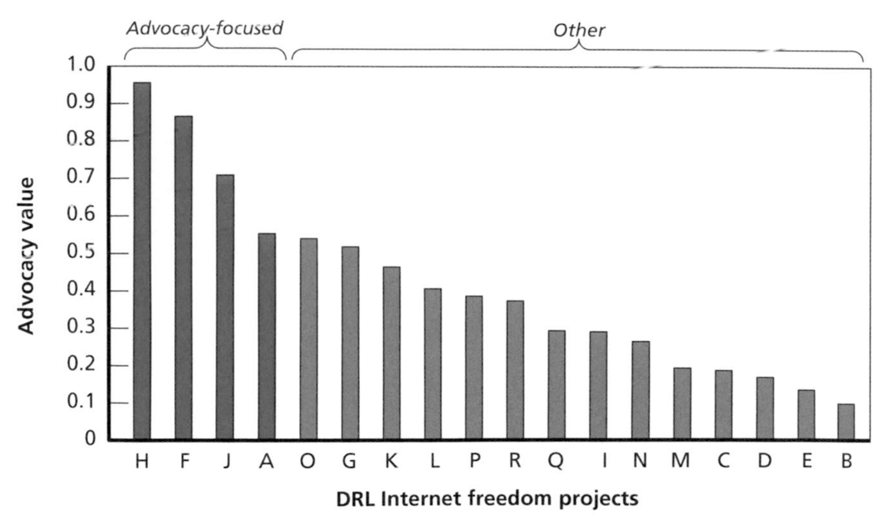

NOTE: Lettered project codes are used to maintain project anonymity.
RAND *RR794-3.4*

Finally, while some of the advocacy projects were active in multiple countries and regions, we observed that some of the strongest projects took a more focused approach, specializing in a single country and employing a multipronged effort to address issues of advocacy from several directions.

Effect of Strategy on Performance

Assessing the individual Internet freedom factors within the portfolio—the access, anonymity, awareness, and advocacy contributions discussed in Chapter Two—provides one perspective of the portfolio's value. We were also interested in assessing the alignment between DRL's Internet freedom strategy and the performance of its funded portfolio of projects. This alignment was captured by looking at the cumulative aspects of these four factors, as well as risk and cost, as discussed. DRL was interested in balancing the contribution from all four factors but was also aware of the critical need for protecting at-risk users within authoritarian regimes.

Strategies should change over time in response to evolving circumstances, and this has proven to be true with the DRL Internet freedom portfolio. Initially, DRL focused on improving access to the Internet in repressive states by funding circumvention tools, such as VPNs and proxy routers. By the time that we began our assessment, DRL had determined that access by itself was insufficient for realizing the goal of expanded political space; users also needed their access to be anonymous and their communications to be secure. Consequently, DRL instructed us to double the value of anonymity in our performance equation.[4] Around the same time, DRL decided that advocacy was less of a priority than the other variables, so its value was halved.[5] As shown in Figure 3.5, the relative weighting of Internet freedom factors, as they

[4] Meeting with DRL Internet freedom team, January 10, 2012, Department of State, Washington, D.C.

[5] Meeting with DRL Internet freedom team, November 14, 2011, Department of State, Washington, D.C.

Figure 3.5
Effect of Different Formulas on Performance

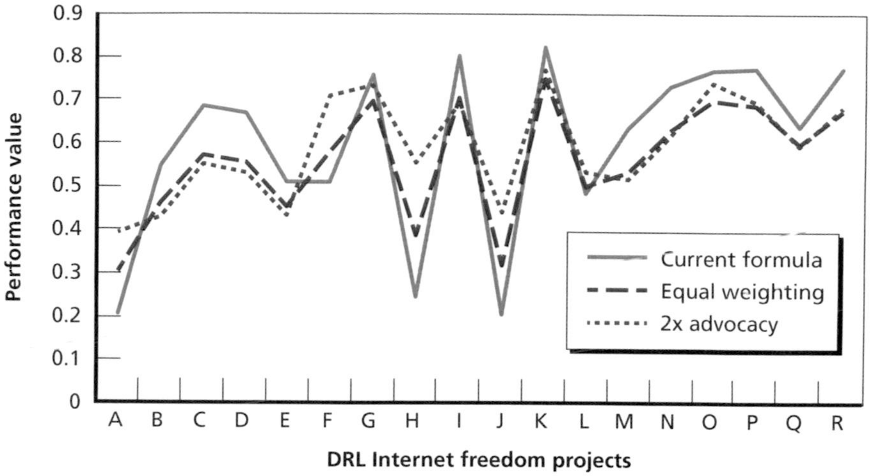

NOTE: Lettered project codes are used to maintain project anonymity.
RAND *RR794-3.5*

reflect the DRL strategy, can normally have a 10–20 percent influence on a project's score. As an excursion, we varied the relative weighting of the advocacy variable from that of the current DRL strategy to that of equal value with the others, as well as to a value twice that of the other factors. Such variations had a notable but contained effect, except when projects were very narrowly focused (e.g., projects H and J); in these cases, the influence exceeded 100 percent.

DRL's Internet freedom strategy is likely to change again. Increasingly, there is recognition that no technical solution alone can ensure a free and open Internet. Instead, the future of the Internet is also affected by legislatures, judges, and international organizations. Consequently, one would expect that advocacy may become a more important part of DRL's Internet freedom strategy. That is not to suggest that previous strategies were flawed; rather, they may have been appropriate for their time, but circumstances have since changed. One of the key benefits of our assessment methodology is that it can adapt to different strategies. If it is employed over time, it can help to ensure that DRL's strategy remains aligned with its funding decisions.

CHAPTER FOUR

Portfolio Balance and Synergy

In addition to assessing the DRL Internet freedom project portfolio's performance, RAND set out to assess the portfolio's balance and synergy. The results revealed that the portfolio was balanced with respect to project focus and geographical distribution, among other factors, and that the program's total effect is greatly enhanced when implementers interact and collaborate.

Balance

In a broad sense, our assessment of the overall portfolio found a strong diversity of effort and balance across the four variables that affect Internet freedom and political space (access, anonymity, awareness, and advocacy). We found these factors were distributed across projects roughly proportionately to their weight in the portfolio performance calculation. Each was addressed by more than half of the projects, which added robustness to the portfolio.

We also found that while the projects exhibited diversity in their scope, objective, and approach, they were generally balanced in their contribution to both overall performance and the four functional variables that were specifically measured. (The two outliers were projects narrowly focused on subcomponents of advocacy). We also determined that the portfolio contained a mix of approaches that were high-risk and high-gain and those that were tried and true. While a handful of projects employed very similar objectives or approaches, these appeared appropriate to the scope of those projects and the desire to have redundancy in that segment of the portfolio. We also found that the diversity along numerous portfolio dimensions was desirable from a portfolio risk-reduction perspective. The portfolio, while balanced, diversified, and distributed among 18 projects, maintained both cohesion with DRL's strategy and a collective unity of effort to be of continued value to targeted users and diplomatic American interests.

Due to this balance and dispersion of Internet freedom factors among DRL projects, we found it helpful to employ an analytic filter to differentiate the projects. After testing several constructs, we found it most helpful and natural to group projects based on their objectives or their functional area. Figure 4.1 displays a PortMan plotting of DRL's Internet freedom projects, with value on the vertical axis and risk on the horizontal axis. In this figure, the projects are categorized into five generic categories that capture their functional rather than substantive areas:

- technology development—programs that concentrated on developing new circumvention or anonymity technologies (which largely correlates with access and anonymity)
- training—programs that focused on training at-risk populations to improve their understanding of their online vulnerabilities and good security practices (which largely correlates with awareness)
- advocacy—advocacy projects that aim to support Internet freedom within states and international organizations
- test—test programs that worked toward ensuring that Internet freedom technologies were robust and did not have security flaws that put their users at risk (these programs did not fall neatly into one of the four Internet freedom variables)
- mixed—mixed programs that were multifaceted and had elements of all of the above.

After categorizing each project by its goal and plotting its performance, we found that projects with the same objective tended to score similarly and cluster together. The technology development and training clusters contributed most to the portfolio's value, while advocacy projects contributed the least, which reflects advocacy's lower weight-

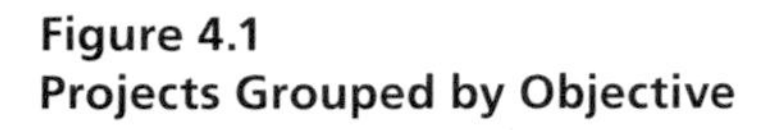

Figure 4.1
Projects Grouped by Objective

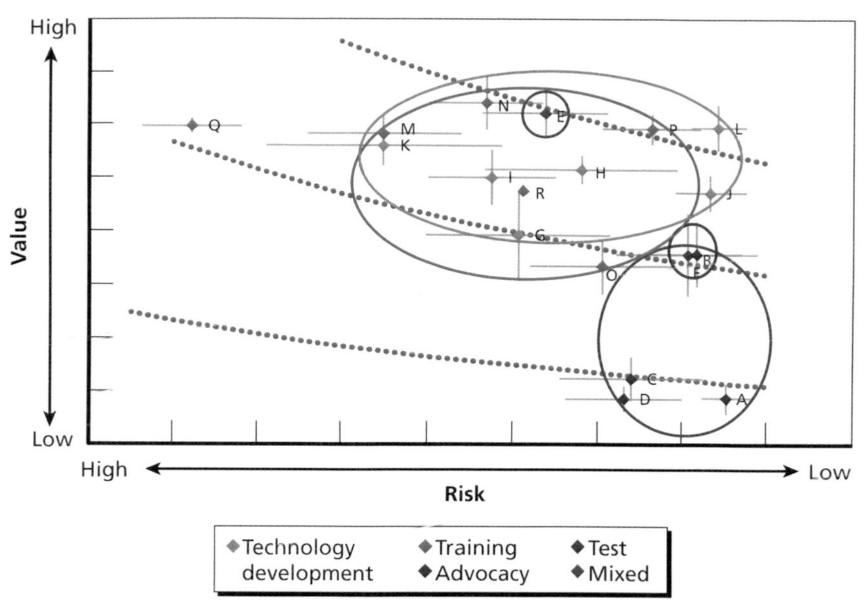

RAND *RR794-4.1*

ing in the performance equation. More interestingly, within each of the three major project clusters, there was balance between high-risk approaches (which hold out the promise of significant gains) and more traditional and proven approaches (which had a higher probability of success but also promised lower returns).

Beyond balance in objectives, approaches, and Internet freedom factors, the portfolio exhibited other characteristics of balance with respect to project investment allocation, geopolitical focus, and breadth. While all the funded projects were clearly aligned with at least one of the key factors influencing Internet freedom and political space, they offered a variety of ways to meet those objectives. Moreover, the investment allocations were fairly evenly apportioned to projects of differing size, ranging from 1 percent to 13 percent of the portfolio's total value (see Figure 4.2). In general, the projects with more resources tended to address multiple aspects of Internet freedom.

Figure 4.2
Division of Funds, by Project

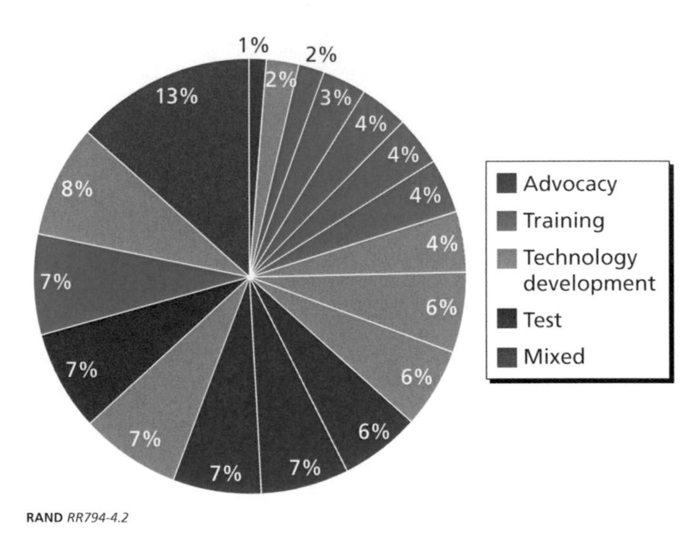

RAND *RR794-4.2*

The geographical reach and geopolitical focus of the projects spanned from global to single country, with an emphasis on regions and countries of particular interest to the USG (e.g., Iran, China, and the Middle East and North Africa) (see Figure 4.3). The most prominent countries fell into two categories: (1) those with high levels of online censorship and surveillance and (2) those experiencing significant internal turmoil (e.g., the Arab spring). This reflected DRL's strategy of focusing on countries that lacked Internet freedom or places where enhancing secure access to the Internet could have a significant effect on expanding political space during a pivotal time. Because the future is difficult to forecast, DRL ensured that it also had some projects that had either a global or regional focus. Five of the projects were targeted at global Internet users, eight had a regional focus, and five had single-country focus.

Finally, the breadth of individual projects within the portfolio was also diverse. While six projects were principally focused on a single Internet freedom factor (access, anonymity, awareness, or advocacy), five spread their focus between two of the factors, another five applied their efforts to three of the factors, and two addressed all four with a

Figure 4.3
Geographic Focus of Projects

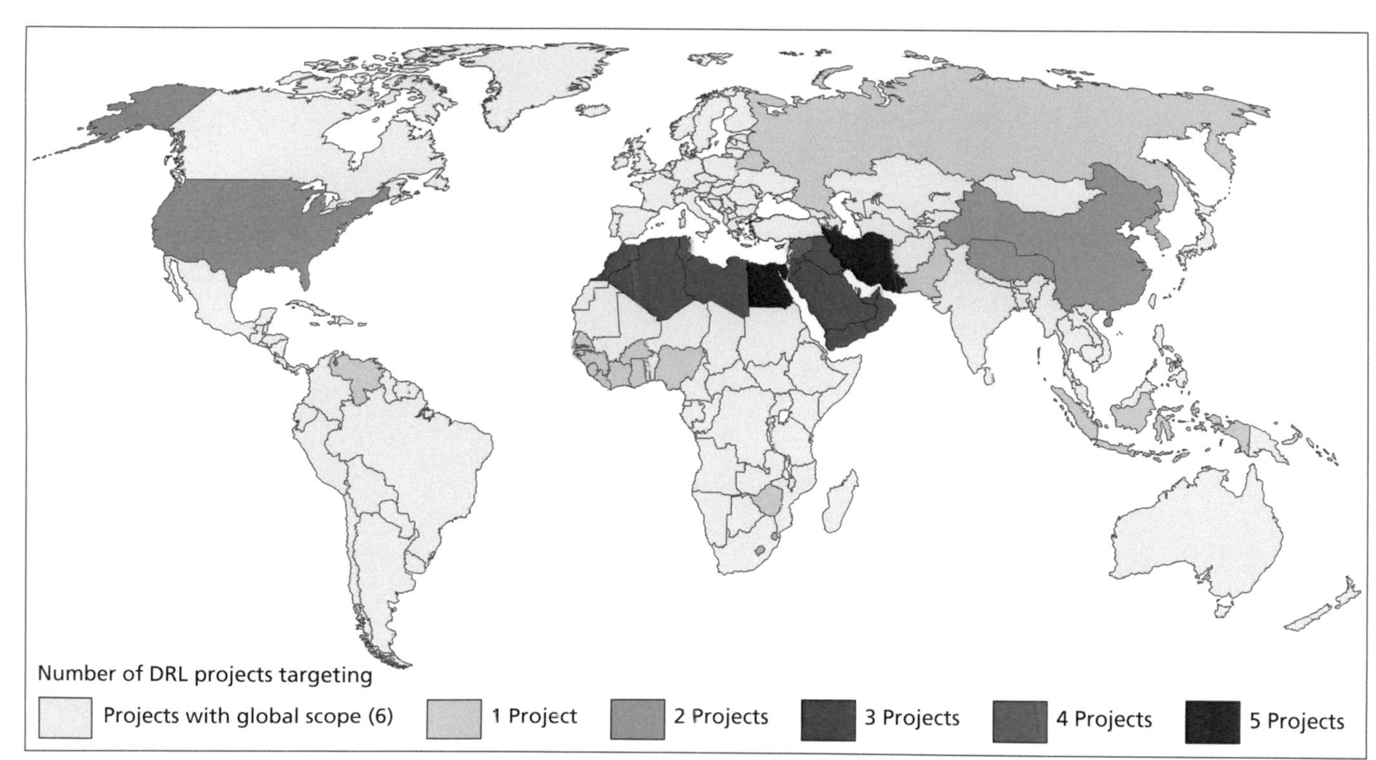

roughly equal level of effort (see Figure 4.4). This is reflective of the diversity of project approaches and the effective distribution of portfolio resources across DRL's Internet freedom agenda.

Moreover, we found that projects were more likely to focus on certain combinations of Internet freedom factors. Projects that had high access scores also tended to have strong anonymity scores. This is because of the growing recognition among developers that tools that provide access also need to offer privacy. There was also a positive and direct—although weaker—relationship between access and awareness (see Figure 4.5). This relationship reflected the fact that many training programs that concentrated on awareness also distributed circumvention technologies to their trainees. As noted earlier, the one exception to this generally positive relationship among Internet freedom factors was advocacy. Programs that scored high in advocacy tended to be more narrowly focused and, therefore, to have lower scores across the other Internet freedom factors.

Figure 4.4
Project Breadth

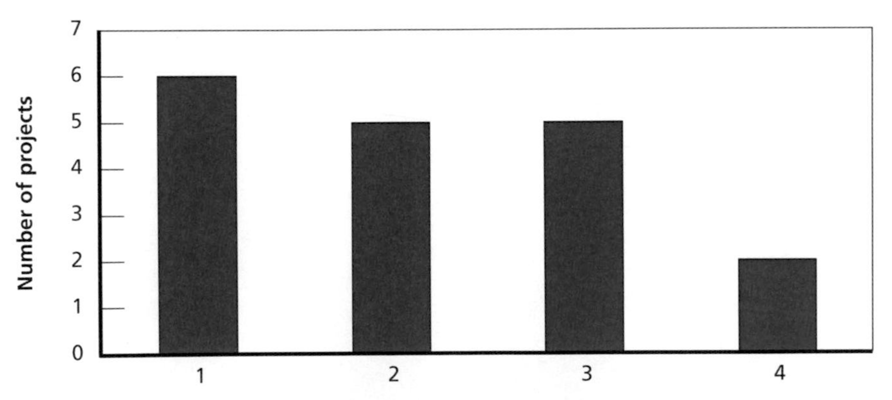

NOTE: Investment allocations ranged from 1 percent to 13 percent of the portfolio's total value.

RAND *RR794-4.4*

Figure 4.5
Comparison of the Value of Internet Freedom Factor Pairs

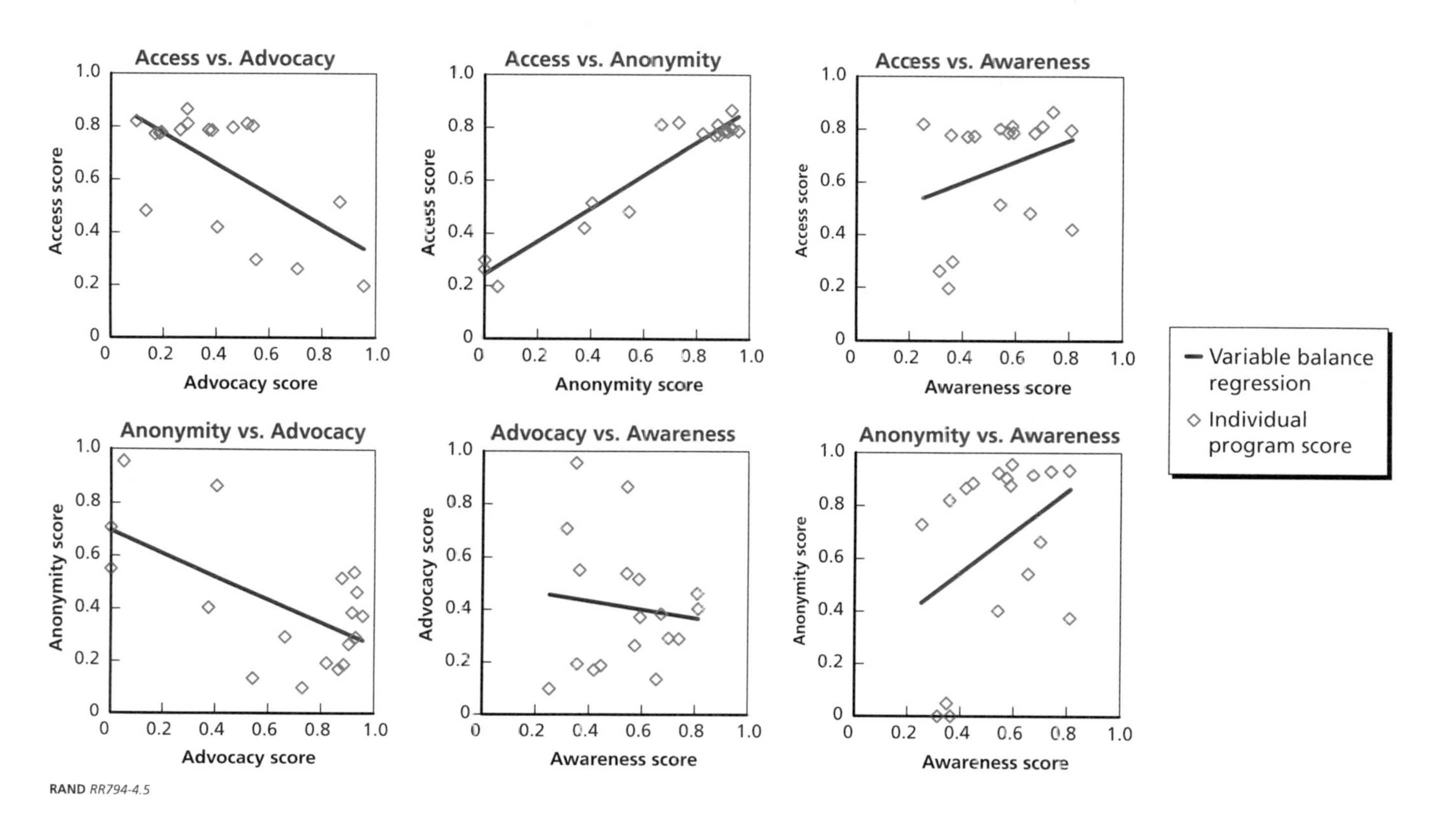

Risk

Besides assessing the DRL portfolio's balance, RAND also assessed potential portfolio *risk*, which we defined as the probability that a project would be successfully implemented.[1] We evaluated three types of risk: capability, or the soundness of a project's approach and staffing; acceptance, or the likelihood that intended users would willingly adopt the project's offering; and sustainability, or the probability that the project would be able to carry on beyond the term of the DRL grant.[2] We then assessed individual projects against these risks to assess their effect on the cumulative portfolio risk.[3]

Figure 4.6 depicts the value of each project along the vertical axis and the level of risk along the horizontal axis. This chart also includes the uncertainty scores, which for risk reflect the amount of disagreement among the SMEs. Projects in the upper right corner are the most desirable because they are the most likely to meet their objectives during the grant and to continue to provide value in the future. Projects falling within the lower right quadrant are likely to be successfully executed, but also contribute less to the portfolio's overall value. Yet these low-risk, low-value projects are balanced by the high-value, high-risk projects found in the upper left quadrant. No projects fell into the lower left quadrant, which is the least desirable area because these

---

[1] At the time of this assessment, projects were in various stages of execution. Investigation of risk was based on material available at the time and the competency and track record of project staff to successfully execute their objectives. In agreement with DRL, is was understood that this assessment could only reasonably investigate near-term risk, and that mid- and far-term risk were out of the assessment's scope.

[2] These projects were in various stages. Two projects were just beginning and 16 were in the middle of executing their tasks when they were interviewed. As a result, the capability risk measure differentiated between proven capability risk (or achievements thus far) and prognostic capability risk (or solidity of approach, staffing, etc.). The total capability risk score, therefore, was calculated based on how far along the project was. For instance, if a project was complete, its capability risk was composed entirely of proven measures, and if a project was only partially executed, its score would be half proven, half prognostic.

[3] Technical and abuse risk were beyond the scope of this assessment. *Technical risk* includes the project's potential to fall short of its stated capability thresholds. *Abuse risk* addresses the potential for third parties to use the project, or its products, for illicit or undesirable purposes. Subsequent to the assessment covered by this report, DRL engaged RAND to assess technical and abuse risk, which will be the subject of a forthcoming report.

Figure 4.6
Project Risk and Value, Including Uncertainty

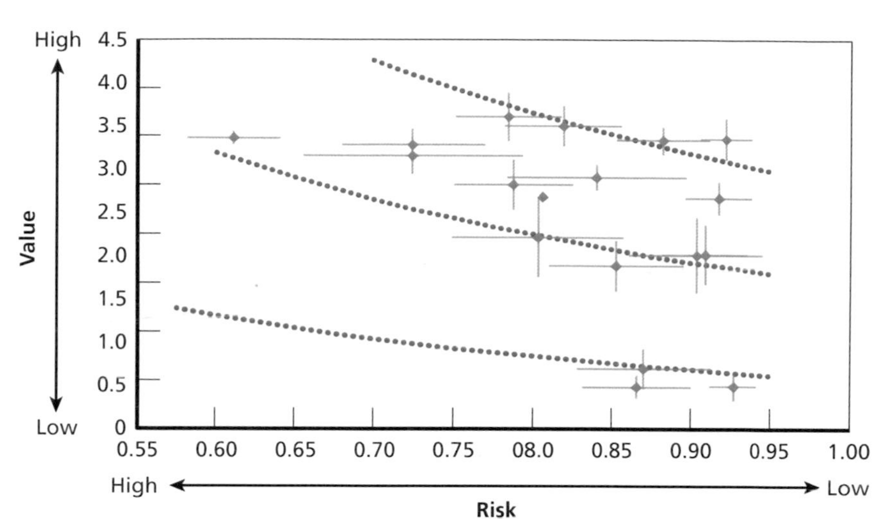

RAND *RR794-4.6*

projects would be both low value and high risk. Moreover, when taking uncertainty into account, most of the projects had low risk scores (that is, above 0.75), which balances out those few high-risk, high-reward projects in the upper left corner.

In short, DRL's portfolio is characterized by an acceptable amount of risk tolerance. If DRL chose to minimize risk, it would also need to scale down projects' objectives and value, which, in turn, would likely decrease the portfolio's overall value. We determined that it was best for DRL to balance its portfolio between some high-risk, high-reward grants and a larger number of low-risk, lower-reward projects. If DRL were to adopt a risk-intolerant approach, in all likelihood, it would sacrifice significant value. Therefore, we concluded that DRL has effectively managed risk versus reward in its Internet freedom portfolio.

Costs

In addition to balance and risk, we assessed the costs associated with DRL's Internet freedom portfolio. Cost had several components, including direct cost (the project's level of funding) and indirect costs (such as domestic or international political costs). As discussed, we found that direct costs were relatively evenly divided among the grantees (see Figure 4.2), with the better-resourced projects generally focusing on multiple aspects of Internet freedom.

Political cost assessed the likelihood that a project might generate negative diplomatic, domestic political, or media effects. Under political cost, we considered the project's geographic focal point, the degree of real and potential interest of the host government in its activities, the project's public profile, ethical and security standards, the organization's attitude toward working with the media, and its relationships with various government entities. Also as a part of political cost, we explored the cost to a program's intended users—particularly whether their security was jeopardized by participating in this program or using an implementer's product and what steps a project took to mitigate these vulnerabilities.

Across the DRL suite of projects, we found that the assessed political costs had yet to materialize.[4] That does not mean that problems might not arise in the future. In working with the various projects, RAND found each to be sensitive to those costs, especially the security of their intended users. Beyond complying with USG guidance, project personnel appeared to feel a deep-seated personal responsibility to do everything possible to ensure the security of their users.

Tor

Despite the diversity and balance within DRL's portfolio, we noted one commonality among several projects: Many relied on Tor, an anonymity tool that also enables one to circumvent Internet filtering. Tor is a multiple-hop proxy router that works by routing Internet traffic through several proxies, thereby bypassing firewalls and protecting the user's identity. While Tor does receive USG funds, it is not a direct DRL grantee and, therefore, was not directly assessed as a project in

---

[4] The majority of projects were evaluated in their early stages, so in those cases, we projected the management and program risk for their assessments.

the portfolio.[5] But because half of DRL's projects used Tor to varying degrees, its influence on the portfolio was of interest.[6] We found that Tor had a positive but not determinate effect on those projects that included it in their approach (see Figure 4.7). The degree of impact was a function of the project's reliance on Tor to protect its users' identity and provide them with a means of accessing blocked websites, which influenced a project's access and anonymity performance scores. An additional unexpected benefit from Tor was the positive contribution it made to the portfolio's synergy.

Figure 4.7
Effect of Tor on Project Performance

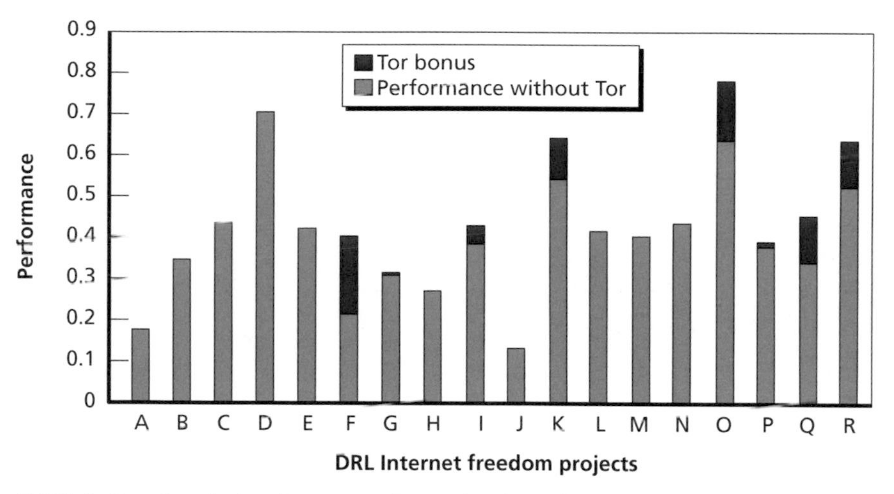

NOTE: Lettered project codes are used to maintain project anonymity.
RAND *RR794-4.7*

---

[5] Roger Dingledine, Nick Mathewson, and Paul Syverson, "Deploying Low-Latency Anonymity: Design Challenges and Social Factors," *Security & Privacy, IEEE*, Vol. 5, No. 5, September/October 2007.

[6] Five of the projects had individuals affiliated with Tor as members of their technology development teams.

Synergy

A key observation of the portfolio assessment was that the program's total effect on Internet freedom is greatly enhanced by interaction and collaboration among implementers. Projects from the five cluster areas (technology development, training, technology testing, advocacy, and mixed efforts) intersect within the portfolio and produce opportunities for project synergy that can lead to enhanced project, and overall portfolio, performance. Such links could also provide conduits for additional collaboration beyond the scope and time frame of the DRL grant. The potential benefit of this synergy is substantial. For example, projects engaging in technology development could greatly benefit from interacting with groups testing for security flaws; training programs could distribute newly developed circumvention tools and tailor them for a particular setting, and so on. Nurturing and enhancing this synergy would provide DRL with the opportunity to notably leverage both the effectiveness and influence of its Internet freedom investment, while reducing portfolio risk.

While we did not develop a formal methodology to assess synergy within the DRL portfolio, the interviews provided insight into where opportunities exist for enhancing synergy. One standout observation was that, from an individual project perspective, generating synergy was often challenging because many projects did not approach their work with a collaborative mindset. There was also tension due to the competition for scarce resources. As a result, we found that there was a notable difference between the actual synergy effects and the potential for creating stronger synergy among portfolio projects.[7]

To better understand the portfolio's potential synergy, RAND researchers mapped the existing connections between projects, which resulted in the network depicted in Figure 4.8. Some implementers had strong, established connections with other groups (shown as large circles), but these were often based on preexisting connections and relationships. Nevertheless, these ties were beneficial because projects

---

[7] These observations were briefed to the DRL portfolio management team during the course of the RAND assessment, and by the assessment's conclusion, DRL had adopted several informal RAND recommendations regarding portfolio synergy.

Figure 4.8
Connections Among Projects

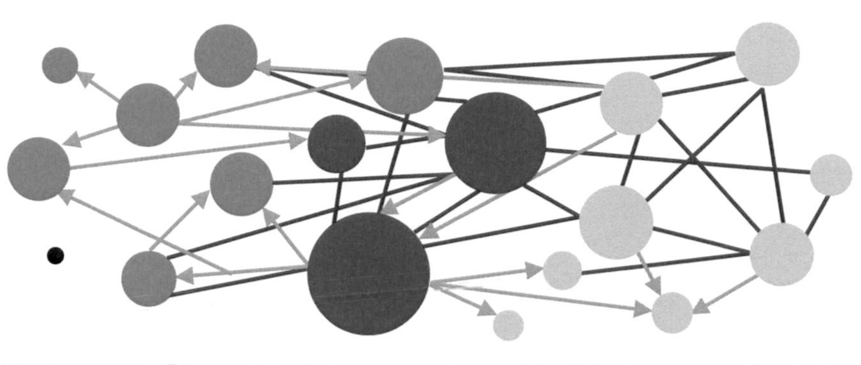

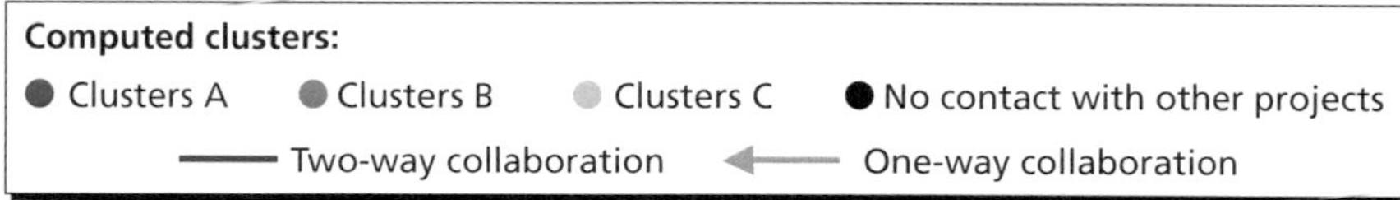

NOTE: Circle size depicts relative connectivity.
RAND *RR794-4.8*

shared best practices and technologies and helped each other overcome problems. Other implementers had fewer ties to other groups (shown as proportionately small circles), and one group was isolated, with no connections to other DRL grantees (shown as a disconnected black dot). Some connections were one-way (shown in gray) while some were more equitable or two-way (shown in red). As discussed, Tor produced an added element of synergy by providing capabilities to several projects. When we applied a commercial network analysis tool to our recorded observations, it produced the characterization of the portfolio depicted in the figure.[8] This process revealed that there were three principal clusters of collaboration, which is similar to what we informally observed. Interviews with project leaders suggested that these clusters were mainly the result of previous working relationships among personnel from different projects that carried over to the DRL portfolio.

---

[8] We used Gephi, an open source interactive and visualization software application for network analysis. See Gephi, *The Open Graph Viz Platform*, 2012.

While there is considerable value to the existing ad hoc connections, we determined that more networking would help to realize the portfolio's full potential. Before our assessment had even concluded, DRL heeded our advice and had taken a number of steps to encourage collaboration among its grantees, including setting up mechanisms to facilitate communication and holding gatherings so that grantees can meet and interact with each other in person. DRL convenes an annual implementers meeting, which is useful for making introductions, learning about other projects, and building trust. Additionally, DRL encouraged the implementers to create a listserv so that they could reach out directly to each other. These are important steps that encourage DRL implementers to seek advice from each other and share best practices, to leverage each other's contacts, and even to cost-share administrative help, but we believe that DRL should do more to foster these connections. To overcome the tension that exists between competition and collaboration, DRL may need to consider incentivizing interproject cooperation by making it one of the criteria in future solicitations or explicitly incorporating it in the grants.

Based on our understanding of the DRL portfolio and its objectives, we constructed an idealized characterization of the portfolio's potential synergy, shown in Figure 4.9. The five cluster areas (technology development, training, advocacy, testing, and mixed efforts) served as the synergy building blocks. In attempting to maximize the portfolio's synergy, there would be an inherent tension among projects cooperating and competing in DRL's limited resource environment. Therefore, collaboration within clusters (i.e., projects that have similar goals), where competition might be highest, would need to be incentivized by DRL. Internal cluster cooperation would involve sharing best practices and lessons learned. Similarly, there is potentially great value to implementers cooperating outside of their own cluster because there is a natural division of labor when pursuing different objectives. Nevertheless, DRL may need to broker intercluster cooperation, which often does not materialize because it goes beyond the scope of the grant. For example, technology developers need agents to distribute their tools, and trainers need Internet freedom tools to protect and empower their

Figure 4.9
Idealized Connections Among Projects

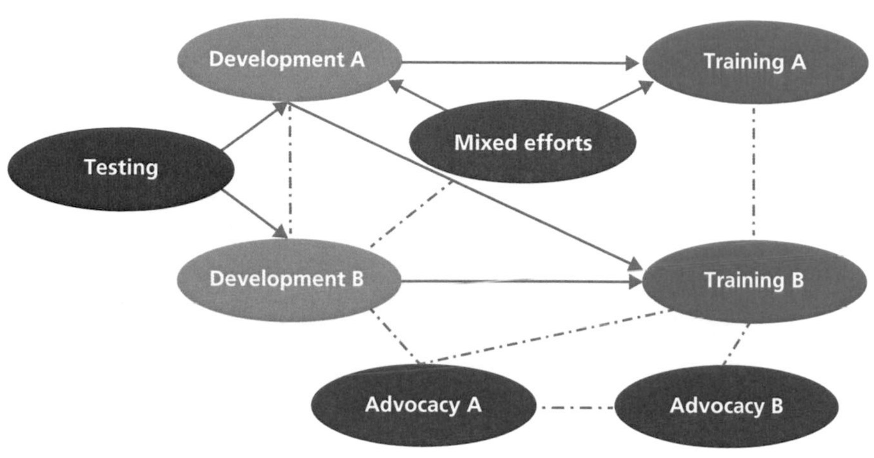

RAND *RR794-4.9*

students; therefore, DRL could benefit both the portfolio and the individual projects by facilitating their partnering.

Ideally, one would want to foster collaboration across USG-funded Internet freedom projects, which would include those who receive grants from the Broadcasting Board of Governors (BBG), DRL, Defense Advanced Research Projects Agency (DARPA), and United States Agency for International Development (USAID). Given how important we found synergy to be for the DRL portfolio, there are potentially significant gains to be realized by doing so. In fact, we did find that some interagency cooperation does exist. For instance, in the past, DRL has invited USAID Internet freedom grantees to its annual gathering. Additionally, individuals from DARPA and other USG agencies sit on the DRL Internet freedom selection committee. But systematically and comprehensively assessing crossdepartmental Internet freedom synergy was beyond the scope of this project, which was focused only on DRL. Moreover, because interagency cooperation and coordination is often quite difficult, we focused on assessing synergy among the limited group of grantees that DRL can directly influence—those within its own portfolio. Promoting connections among the broader USG Internet freedom community merits further study and consideration.

Because the potential for cooperation was not fully realized, we argue that synergy is one area that could be improved in notable ways. Synergy presents an especially worthwhile investment because of its low cost and high potential payoff. More than just increasing the portfolio's effectiveness and efficiency, enhanced synergy could provide other positive latent effects. Even if projects do not immediately collaborate from start to finish, the latent relationships facilitated by their interacting with other DRL participants may bear fruit at a later time. In particular, fostering personal and organizational ties and enhancing trust would position the community to organically and swiftly respond to rapidly developing crises associated with Internet freedom.

CHAPTER FIVE

Additional Observations

Technology Development

Developing new technologies that enable individuals to have unfettered and secure access to the Internet is a significant but complicated goal of the DRL portfolio. The struggle between those promoting Internet freedom and those trying to control and censor the Internet is a fast-paced game of cat and mouse. Consequently, the speed of this contest often outstrips the grant cycle, and implementers often have to modify their proposed deliverables in response to developments on the ground. Responding effectively to the countermoves made by authoritarian governments is difficult under any circumstances. This predicament is further complicated by the fact that technology development is not a traditional State Department activity and, therefore, not one of its core capabilities. Partnering with other USG entities that have proven technology development infrastructure and core competencies, such as DARPA, could help to overcome this limitation.[1]

At the broadest level, the State Department is interested in nurturing two types of technological advancements: making evolution-

[1] DARPA, which also has an Internet freedom technology development program, participates in selecting DRL projects to include in the portfolio. Based on interviews with the DARPA and DRL Program Managers, we found that there is an informal understanding between DARPA and DRL that DARPA will tackle longer-term and technology-difficult challenges, and DRL will focus its portfolio on nearer-term challenges. Because, at the time of this assessment, there was little commercial market demand for increased security for non-institutional users in their everyday communications, there was a paucity of for-profit firms with which DRL could work.

ary improvements to existing circumvention and anonymity tools and incubating new revolutionary Internet freedom capabilities. DRL's current portfolio has a mix of improving known and proven tools (or tailoring them to particular countries) and supporting the development of new and paradigm-shifting capabilities.[2] Doing so effectively, however, requires clear, measurable performance targets for the grantees and the use of trusted third parties (e.g., an independent lab) to test and validate both types of tools.

DRL's investment in Internet freedom should have enduring value that extends well beyond the life of the individual grants. In particular, one of the most important effects of the DRL portfolio appears to be the community it nurtures. Although more might be done, DRL has taken steps to maximize the return on its Internet freedom portfolio by bringing together the individuals, organizations, and tools it supports. The DRL Internet freedom community is only one subset of the larger Internet freedom community, which includes organizations supported by other U.S. agencies (including DARPA, USAID, the BBG, and perhaps other unacknowledged organizations), foreign governments, nongovernmental organizations (NGOs), and private citizens and corporations, such as Google.[3]

The DRL Internet freedom community could play different roles, depending on the situation. During noncritical times, the DRL Internet freedom community is focused on increasing the State Department's steady-state capability to promote freedom online. As discussed, it does so by encouraging formal and informal collaboration among grantees to improve the efficacy of both projects. At the same time, fostering these ties also develops latent surge capacity to respond during Internet freedom crises. At critical moments, the DRL-sponsored community has the capacity to rapidly and independently respond to developments in an effort to expand political space. In these circumstances, a

---

[2] Based on the above DRL-DARPA understanding, the technology development aspects of DRL's strategy were to support low-risk, country-targeted, global, and uniquely innovative technologies that were not being adequately addressed elsewhere in government or industry.

[3] For more on USG agencies involved in Internet freedom, see Figliola, 2011, and Google, *Google Take Action*, undated.

self-synchronizing community composed of independent actors whose interests are aligned with the USG is not only well-positioned to react, given its grassroots connections, but also more agile and capable of responding in a timely fashion than the government. Moreover, having the USG stay in the background reduces the potential for blowback. In short, part of the DRL portfolio's key lasting value is in the relationships it builds and the capability of those relationships to generate a rapid, coherent response to emergent Internet freedom crises.

Characteristics of Well-Run Projects

From the interviews, we identified five key characteristics for a well-run project—which have a strong correlation to successfully managing portfolio risk. The first three of these aspects are dependent on the project's human capital: (1) the person(s) who conceptualized the project or continued to serve as its champion within the project's larger host organization; (2) the team member(s) with the functional expertise that ensured that the project had a sound technical or methodological approach, and (3) the project manager(s) who ensured that there is a viable roadmap for achieving the project's objective and being on budget and on schedule in its execution. In the strongest projects, three different individuals separately handled these areas, though there were notable exceptions when one individual simultaneously and successfully filled two roles. When one individual attempted to fill all three roles, some aspect of the project appeared to suffer from inattention.

The fourth aspect of a well-run project revealed by the interviews was its links to other projects, pockets of expertise, and insight outside of the particular project. Those projects that interacted with knowledgeable external actors appeared to use the insights gained to fine-tune their approach and help them navigate roadblocks encountered during execution. Those projects that specifically interacted with others in the DRL portfolio also appeared more comfortable in their DRL working relationship and more knowledgeable.

The final aspect of a well-run project, the program environment, addressed the circumstances in which the project was executed. The more benign the program environment, the lower the stress on the project and the higher the probability of achieving its objectives. The program environment had three elements: the project's host organization, the project's working relationship with DRL, and where the project was being executed. We found that projects had a spectrum of host organizations, from large universities to essentially single-person operations, with large, medium, and small nonprofits in between. We observed no correlation between the nature of the host organization and the performance of projects. This is likely a result of DRL's grant screening process, which filters for quality. From the project's perspective, we observed a consistently positive working relationship with DRL across the portfolio.[4] Finally, there was much diversity in where projects were being executed. Some were in the most authoritarian of states with high personal risk to participants, some in locations with only rudimentary information technology (IT) infrastructure, and some in high-tech test environments. We again viewed this portfolio diversity as an indication of its robustness.

The specific criticality of these factors—the distribution of leadership workload among the champion, the functional expert, and project management and the desirability of strong external links—were not part of our formal assessment. They were observed and learned during the course of the assessment and therefore were not part of the assessment's formal methodology. Nonetheless, we found a strong informal correlation between their presence and low management and performance risk. They also correlated with projects staffed by more experienced personnel.

Common Challenges

While we assessed numerous strengths in the DRL Internet freedom portfolio, there were some underlying areas of potential concern,

[4] These observations were made in the early phases of each project. RAND has no data on the working relationship beyond that observation period. Those organizations who had worked with DRL previously through a project's full lifecycle indicated that, based on past experience, they expected no changes in the DRL working relationship.

many of which are outside of DRL's control.[5] The most prevalent and acute problem is one of staffing. Across the portfolio, projects' top tier concern was access to high-quality IT talent. As mentioned, Internet freedom is playing a game of cat and mouse between those who seek to provide all of the benefits of the Internet to all of the world's citizens and those regimes who seek to repress the Internet to enhance their regime's stability. These authoritarian regimes apply their best available IT talent to enhancing their own security by repressing their citizens' Internet activities in one form or another. On the Internet freedom side, this work is principally done by nonprofits and NGOs. To effectively challenge repressive regimes, NGOs need highly talented IT professionals but cannot effectively compete for these individuals in the commercial marketplace, where employment choice is primarily a matter of monetary compensation. Rather, NGOs are limited to a very small subset of the larger market, in which IT experts are motivated more by mission than money. While there are some truly inspiring IT experts within that subset, their supply is insufficient to meet the demand.

Another concern is that some projects are narrowly focused in scope and hold little promise of being scaled up successfully. It seems as if this constrained focus was not always due to resources alone, but rather to the approach that the project had taken. At times, a small group of opinion leaders may have a disproportionate influence on political space, and it may therefore be wise to target this select group. However, it is not always the case that this narrow focus is wise, and, when possible, it is often valuable to expand pilot projects.

Three additional matters are of lesser, but still notable, concern. First, because Internet freedom is a relatively new field for both liberal democratic governments and NGOs, some organizations are starting from scratch while others have only a few years of project experience. Until a body of best practices and lessons learned are codified and accessible to implementers, many will find themselves expending effort on problems previously solved by others. Second, many of the

---

[5] These concerns were those noted as of April 2013; subsequent international events and USG actions may have affected their significance.

project implementers would like to better support the multistakeholder approach to Internet governance, but they are not certain how to do so given the many actors and institutions involved. And third, the nature of the DRL grant process reasonably requires projects to commit to a very concrete set of objectives and a relatively long-term course of action to achieve them. Regrettably, the current nature of the Internet freedom space is punctuated by crises (e.g., the Arab Spring and Syria). At times, these crises create situations where project resources could have more influence if they were diverted from stated plans to responding to the immediate situation on the ground.

Relationship Between Internet Freedom and Political Space

There is a positive, but indirect, connection between DRL's Internet freedom portfolio and the expansion of civic freedom within authoritarian regimes.[6] RAND's analysis of five case studies revealed that the effect that Internet freedom has on political space is dependent on the level of Internet penetration, the reach of the Internet freedom programs, and the regime's repressive capacity. In general, Internet freedom can facilitate social mobilization, but its influence is mediated by a state's regime type. In particular, there is a direct relationship between the degree of repression and the likelihood of popular mobilization. Although the Internet may encourage democratic popular uprisings by increasing the availability of information and creating broad decentralized movements whose members share a common identity, it has also enhanced the ability of states to monitor dissidents and to spread disinformation and propaganda. The effect of the Internet on political space, therefore, is varied and not always positive. Moreover, while greater access to the Internet can trigger democratization or help to oust an autocrat, it does not appear to have as strong of an effect on democratic consolidation (i.e., the creation of a stable democratic regime).

---

[6] This section is drawn from Tkacheva et al., 2013. For more on the relationship between Internet freedom and democracy, see Fontaine and Rogers, 2011, pp. 14–19.

Yet it is also worth noting that nonpolitical uses of the Internet can transition into politicized online mobilization. In other words, expanding online social space can at times inadvertently lead to greater political space. This suggests that promoting Internet freedom, regardless of whether it helps activists, can ultimately result in greater political space. DRL's projects seek to ensure that all people have free access to the Internet, which could be a critical enabler that helps empower opposition movements in repressive states.

While access to the Internet has not proven to be the primary causal factor in successful revolutions, it has played a *critical enabling role* by reducing the costs of collective action, promoting a shared identity among broad, crosscutting coalitions, and spreading information within a society. Given that the Internet greatly reduces the traditional barriers of distance and time, it facilitates the sharing of information. Therefore, at times, Internet access has destabilized authoritarian regimes by generating information cascades, which in turn have drawn a greater number of protesters to demonstrations. Access to the Internet also enables opponents to challenge a government's version of an event by creating and disseminating alternative interpretations that may resonate with others. In sum, the ability to anonymously express dissent online and coordinate the actions of a group can help to create new opposition movements and expand the appeal of existing dissident groups.

Trade-offs

Several trade-offs need to be considered when funding Internet freedom projects. First, one can seek to expand political space in different ways. On the one hand, increasing the absolute number of individuals who choose to circumvent censorship would result in a net increase in political space. On the other hand, Internet freedom programs could focus on an elite group of agenda setters—bloggers, online journalists, and opposition leaders—by providing them with and teaching them how to use sophisticated anonymity, circumvention, and communications technologies. In other words, the first strategy prioritizes broadening Internet freedom for all users, while the second strategy chooses to deepen online opposition by assisting a small number of activists. History suggests that, at times, the actions of a small group of committed activists can serve as the catalyst for regime change, and they can have a disproportionate effect on outcomes. Then again, netizens who are not politically motivated and who far outnumber the few committed activists can, under the right circumstances, become politically active and decisively tilt the balance of power against a regime.

Similarly, DRL's Internet freedom program needs to find the right balance between generally improving Internet freedom and providing support to dissidents during a crisis. The former will generally improve people's lives by safeguarding a basic human right, while the latter has the potential to have a more significant and direct effect on political space. Different combinations of strategies may be appropriate in different contexts. In short, there is no one-size-fits-all approach. Instead, DRL needs to consider the particular environment and circumstances when determining how to best advance Internet freedom.

Finally, there are some concerns about the USG's Internet freedom programs. First, there is often a disconnect between those developing the circumvention and anonymity technologies and the societies that they hope will utilize these tools. Consequently, DRL needs to help bridge this divide by encouraging cooperation between NGOs or country experts and the programmers developing the technologies as a way to improve the likelihood that the tools are tailored to and therefore embraced by a particular society. Additionally, there is a chance that the USG could unintentionally discredit local activists by providing funds to them, so DRL must be mindful of this risk and take appropriate steps to avoid such occurrences.

CHAPTER SIX

Findings, Recommendations, and Conclusions

Through this assessment, RAND determined that DRL currently has a balanced and acceptable portfolio of Internet freedom programs. DRL's portfolio contains a balance among the four factors that influence Internet freedom and political space (access, anonymity, awareness, and advocacy), and it aligns with DRL's stated strategy. At the time of our assessment, we found no insurmountable obstacles to executing the portfolio. DRL has already taken steps to enhance synergy among its grantees, and the structural issue of compensation for highly talented IT professionals is largely beyond the influence of DRL, and is therefore not easily remedied.

Moreover, portfolio risk and political costs have not yet materialized, and we did not detect any currently embedded flaws that would drive future risk and costs significantly higher. That does not mean that unforeseen problems might not arise. While individual projects will, in all probability, fall short in one aspect or another, the portfolio should still support DRL's Internet freedom strategy due to its diversification.

DRL provides a level of risk tolerance in assembling its portfolio of projects through a rigorous and competitive selection process.[1] It also diversified its portfolio to mitigate risk. This diversification was evident in its project investment allocation, geopolitical focus, project breadth, differentiation of capabilities to develop, and ambition of project objectives. DRL has maintained a degree of failure tolerance

[1] Grants are competitively selected on quality of project ideas, program planning, ability to achieve objectives, cost-effectiveness, program monitoring and evaluation, and the host institution's record and capacity.

within its portfolio, which ensures that it does not become too risk-averse and stagnant.

We determined that, based on execution at the time of the assessment, the DRL portfolio has promise for making a positive contribution to the State Department's Internet freedom initiative. This contribution could be enhanced by nurturing a community of interest of current, past, and potential future DRL grantees, which could collectively catalog and share lessons learned and best practices and create a clearinghouse for Internet freedom tools. Such a community could also act in U.S. interests during times of crisis when Internet freedom capabilities might act as an accelerant to enlarging political space. The working and trust relationships built around the DRL portfolio might serve as a foundation from which like-minded members could pool or integrate their expertise to rapidly respond to emerging opportunities to overcome Internet repression. Such voluntary efforts might have the agility to operate inside an authoritarian regime's decision cycles and would certainly be faster than formal USG programmatic responses. The high probability of tight alignment between USG interests and those of this community would significantly mitigate the normal political risks associated with such independent actions.

Beyond its primary objective of expanding political space, the DRL Internet freedom portfolio offers the USG a way of imposing costs on authoritarian adversaries. Cost imposition is not a stated objective of DRL's Internet freedom program, but it is an ancillary benefit that could be particularly important in an era of austerity. Cost-imposing strategies take a long-term approach, focus on asymmetries, and implement measures that capitalize on U.S. strengths while exploiting adversary weaknesses.[2] Internet freedom targets repressive states' Achilles' heels—their internal legitimacy.[3] These regimes demonstrate their concerns by devoting more than 50 percent of their security budgets to

---

[2] For more information, see Thomas G. Manhken, ed., *Competitive Strategies for the 21st Century: Theory, History, and Practice*, Stanford, Calif.: Stanford University Press, 2012.

[3] Robert G. Sutter, *Chinese Foreign Relations: Power and Policy Since the Cold War*, Lanham, Md.: Rowman and Littlefield Publishers, 2012, p. 17; Office of the Secretary of Defense, *Annual Report to Congress: Military and Security Developments Involving the People's Republic of China 2013*, U.S. Department of Defense, 2013, p. 17; Keith Crane, Rollie Lal, and Jeffrey

internal security.[4] Because countries like China and Iran view Internet freedom as a critical threat, they are likely to continue to allocate a disproportionate amount of resources to trying to control the Internet as long as there are easy ways for their citizens to circumvent censorship and surveillance. Further, USG-funded Internet freedom portfolios are a somewhat unique cost-imposing capability—one that has a high-confidence alignment with USG public and private interest and low probability of being compromised or exploited.

Consequently, we believe that these second-order and unintended outcomes might be of equal or greater value to the USG compared to the Internet freedom portfolio's intended first-order objectives. In large part, this is because the community of interest and cost-imposing capabilities do not disappear when the current funding cycle ends. Rather they deliver a long-lasting residual value to the USG.

As a result of our assessment, we developed four recommendations for DRL to consider as their portfolio continues to mature. Our first recommendation, which DRL began to address during the course of the assessment, is to enhance the synergy within the portfolio and among its grantees. As discussed, this is the area that would provide the largest return on investment for DRL. It would allow them to increase the effectiveness of the entire portfolio while decreasing its management and performance risk. Beyond merely providing opportunities for intraportfolio collaboration, we also recommended that DRL create mechanisms to incentivize collaboration and that these be outlined in the requests for proposals. Additionally, we encourage DRL and other USG agencies working on Internet freedom to increase collaboration among the broader USG-funded Internet freedom community.

Second, we urge DRL to maintain a relatively balanced Internet freedom strategy that includes projects working on access, anonymity, awareness, and advocacy. While it may be necessary that one or sev-

---

Martini, *Iran's Political Demographic, and Economic Vulnerabilities*, Santa Monica, Calif.: RAND Corporation, MG-693-AF, 2008.

[4] Jeremy Page, "Internal Security Tops Military in China Spending," *Wall Street Journal*, March 5, 2011; Jalil Roshandel, *Iran, Israel, and the United States: Regime Security vs. Political Legitimacy*, Santa Barbara, Calif.: Praeger Security International, 2011.

eral factors are given priority at a particular time, DRL should remain active in all four areas. It is increasingly apparent that circumvention and anonymity technologies alone cannot preserve Internet freedom. Instead, it is important both to train the at-risk individuals in how to use these technologies and to support efforts that uphold a free and open Internet in the domestic and international political arenas.

Third, we recommend that DRL consider a resourcing mechanism for contingency tasking. This recommendation grew from observing that several of the projects were actively seeking ways that they could leverage their DRL funds or their work for DRL to respond to the rapidly deteriorating situation in Syria. This demonstrated that a key value of the DRL portfolio was the residual capability of the network that it helped build. Having a way to rapidly and robustly energize that network to respond to other emerging crises would provide the government with a unique soft-power tool.

Finally, we recommend that DRL consider this assessment as a rigorous first look at its portfolio. To fully realize its value, this process should be repeated over time. This is a one-time assessment, which accurately represents a snapshot of DRL's Internet freedom portfolio in fiscal year 2012–2013. But as we have discussed, circumstances and strategies change—often very rapidly in this arena. Therefore, the Internet freedom portfolio needs to be periodically reassessed to monitor its response to these changes and to ensure that it is still optimized to achieve the State Department's objective of expanded political space.

References

Aday, Sean, Henry Farrell, Marc Lynch, John Sides, and Deen Freelon, *Blogs and Bullets II: New Media and Conflict After the Arab Spring*, Peaceworks No. 80, Washington, D.C.: United States Institute of Peace Press, July 2012.

Aday, Sean, Henry Farrell, Marc Lynch, John Sides, John Kelly, and Ethan Zuckerman, *Blogs and Bullets: New Media in Contentious Politics*, Peaceworks No. 65, Washington, D.C.: United States Institute of Peace Press, September 2010.

Anderson, Collin, "Dimming the Internet: Detecting Throttling as a Mechanism of Censorship in Iran," June 18, 2013. As of February 24, 2014:
http://arxiv.org/pdf/1306.4361v1.pdf

Baer, Daniel, Deputy Assistant Secretary, Bureau of Democracy, Human Rights, and Labor, "Promises We Keep Online: Internet Freedom in the Organisation for Security and Cooperation in Europe (OSCE) Region," Statement Before the Commission on Security and Cooperation in Europe (U.S. Helsinki Commission), Washington, D.C., July 15, 2011. As of February 24, 2014:
http://www.state.gov/j/drl/rls/rm/2011/168684.htm

Chomiak, Laryssa, and John P. Entelis, "The Making of North Africa's Intifada," in David McMurray and Amanda Ufheil-Somers, eds., *The Arab Revolts: Dispatches on Militant Democracy in the Middle East*, Bloomington, Ind.: Indiana University Press, 2013.

Conrad, Edward D., Michael D. Michalisin, and Steven J. Karau, "Measuring Pre-Adoptive Behaviors Toward Individual Willingness to Use IT Innovations," *Journal of Strategic Innovation and Sustainability*, Vol. 8, No. 1, June 2012, pp. 81–92.

Crane, Keith, Rollie Lal, and Jeffrey Martini, *Iran's Political Demographic, and Economic Vulnerabilities*, Santa Monica, Calif.: RAND Corporation, MG-693-AF, 2008. As of February 24, 2014:
http://www.rand.org/pubs/monographs/MG693.html

Dalkey, Norman Crolee, *The Delphi Method: An Experimental Study of Group Behavior*, Santa Monica, Calif.: RAND Corporation, RM-5888-PR, 1969. As of February 24, 2014:
http://www.rand.org/pubs/research_memoranda/RM5888.html

DeNardis, Laura, *The War for Internet Governance*, New Haven, Conn.: Yale University Press, 2014.

Dickinson, Elizabeth, "Internet Freedom: The Prepared Text of U.S. Secretary of State Hillary Rodham Clinton's Speech, Delivered at the Newseum in Washington, D.C.," *Foreign Policy*, January 21, 2010. As of February 24, 2014: http://www.foreignpolicy.com/articles/2010/01/21/internet_freedom

Diebert, Ronald, and Rafal Rohozinski, "Beyond Denial: Introducing Next Generation Information Access Controls," in Ronald Deibert, John Palfrey, Rafal Rohozinski, and Jonathan Zittrain, eds., *Access Controlled: The Shaping Power, Rights, and Rule in Cyberspace*, Cambridge, Mass.: MIT Press, 2010. As of February 24, 2014:
http://mitpress.mit.edu/sites/default/files/titles/content/9780262514354_sch_0001.pdf

Dingledine, Roger, Nick Mathewson, and Paul Syverson, "Deploying Low-Latency Anonymity: Design Challenges and Social Factors," *Security & Privacy, IEEE*, Vol. 5, No. 5, September/October 2007. As of February 24, 2014:
http://ieeexplore.ieee.org/stamp/stamp.jsp?tp=&arnumber=4336287&isnumber=4336264

Elster, Jon, ed., *Rational Choice*, New York: New York University Press, 1986.

Faris, Robert, and Rebekah Heacock, "Introduction," in Urs Gasser, Robert Faris, Rebekah Heacock, *Internet Monitor 2013: Reflections on the Digital World*, Cambridge, Mass.: Berkman Center for Internet and Society, December 12, 2013a.

———, "Looking Ahead," in Urs Gasser, Robert Faris, Rebekah Heacock, *Internet Monitor 2013: Reflections on the Digital World*, Cambridge, Mass.: Berkman Center for Internet and Society, December 12, 2013b, pp. 86–87.

Faris, Robert, and Nart Villeneuve, "Measuring Global Internet Filtering," in Ron Deibert, John Palfrey, Rafal Rohozinski, and Jonathan Zittrain, eds., *Access Denied: The Practice and Policy of Global Internet Filtering*, Cambridge, Mass.: MIT Press, 2008.

Figliola, Patricia Moloney, *Promoting Global Internet Freedom: Policy and Technology*, Washington, D.C.: Congressional Research Service, R41837, May 26, 2011. As of February 24, 2014:
http://fpc.state.gov/documents/organization/166789.pdf

Figliola, Patricia Moloney, Kennon H. Nakamura, Casey L. Addis, and Thomas Lum, *U.S. Initiatives to Promote Global Internet Freedom: Issues, Policy, and Technology*, Washington, D.C.: Congressional Research Service, R41120, January 3, 2011. As of February 24, 2014:
http://www.fas.org/sgp/crs/misc/R41120.pdf

Fontaine, Richard, and Will Rogers, *Internet Freedom: A Foreign Policy Imperative in the Digital Age*, Washington, D.C., Center for New American Security, June 2011.

Galperin, Eva, and Morgan Marquis-Boire, "Syrian Activists Targeted with Facebook Phishing Attack," *Electronic Frontier Foundation*, March 29, 2012. As of February 24, 2014:
https://www.eff.org/deeplinks/2012/03/pro-syrian-government-hackers-target-syrian-activists-facebook-phishing-attack

Gephi, *The Open Graph Viz Platform,* 2012. As of February 24, 2014:
https://gephi.org/

Gladwell, Malcolm, "Small Change: Why the Revolutions Will Not Be Tweeted," *New Yorker,* October 4, 2010. As of July 6, 2014:
http://www.newyorker.com/reporting/2010/10/04/101004fa_fact_gladwell

Google, *Google Take Action*, undated. As of February 24, 2014:
http://www.google.com/intl/en/takeaction/

Grantsolutions.gov, "Bureau of Democracy, Human Rights and Labor and Bureau of Near Eastern Affairs Joint Request for Statements of Interest: Internet Freedom Programs," May 2012. As of August 9, 2014:
https://www.grantsolutions.gov/gs/preaward/previewPublicAnnouncement.do?id=14367&timeZone=EST&cancelMapping=cancelToAnnouncements

Gurr, Ted Robert, *Why Men Rebel: Fortieth Anniversary Edition*, Boulder, Colo.: Paradigm Publishers, 2011.

Hanson, Fergus, *Baked In and Wired: eDiplomacy@State*, Washington, D.C.: Brookings Institution, October 25, 2012.

Harris, Leslie, et al., "An Open Letter to Congress About Internet Freedom," web page, March 14, 2011. As of July 16, 2014:
http://mobileactive.tumblr.com/post/3854828930/letter

Helmer-Hirschberg, Olaf, *Systematic Use of Expert Opinions*, Santa Monica, Calif.: RAND Corporation, P-3721, 1967. As of February 24, 2014:
http://www.rand.org/pubs/papers/P3721.html

Howard, Philip N., and Muzammil M. Hussain, "The Role of Digital Media," *Journal of Democracy,* Vol. 22, No. 3, 2011, pp. 35–48.

———, *Democracy's Fourth Wave: Digital Media and the Arab Spring*, Oxford, UK: Oxford University Press, 2013.

International Telecommunications Union, *The World in 2013: ICT Facts and Figures*, 2013. As of February 24, 2014:
http://www.itu.int/en/ITU-D/Statistics/Documents/facts/ICTFactsFigures2013-e.pdf

Jervis, Robert, *Perception and Misperception in International Politics*, Princeton, N.J.: Princeton University Press, 1976.

Joseph, Sarah, "Social Media, Political Change, and Human Rights," *Boston College International and Comparative Law Review,* Vol. 35, No. 1, January 1, 2012, pp. 145–150.

Kelly, Sanja, "Despite Pushback, Internet Freedom Deteriorates," in Sanja Kelly et al., *Freedom on the Net 2013: A Global Assessment of Internet and Digital Media*, Washington, D.C.: Freedom House, 2013.

Kelly, Sanja, and Sarah Cook, eds., *Freedom on the Net 2011: A Global Assessment of Internet and Digital Media*, Washington, D.C.: Freedom House, April 18, 2011.

Kelly, Sanja, Sarah Cook, and Mai Truong, eds., *Freedom on the Net 2012: Global Assessment of Internet and Digital Media*, Washington, D.C.: Freedom House, 2012.

Kloc, Joe, "Syria Grants Free Internet Access So It Can Snoop," *Newsweek*, February 27, 2014.

Landree, Eric, Richard Silberglitt, Brian G. Chow, Lance Sherry, and Michael S. Tseng, *A Delicate Balance: Portfolio Analysis and Management for Intelligence Information Dissemination Programs,* Santa Monica, Calif.: RAND Corporation, MG-939-NSA, 2009. As of February 24, 2014:
http://www.rand.org/pubs/monographs/MG939.html

Lewis, Alan A., *The Use of Utility in Multiattribute Utility Analysis*, Santa Monica, Calif.: RAND Corporation, P-6396, 1980. As of February 24, 2014:
http://www.rand.org/pubs/papers/P6396.html

Linbach, Mark I., *The Rebel's Dilemma*, Ann Arbor, Mich.: University of Michigan Press, 1998.

Lum, Thomas, Patricia Moloney Figliola, and Matthew C. Weed, *China, Internet Freedom, and U.S. Policy,* Washington, D.C.: Congressional Research Service, R42601, July 13, 2012. As of February 24, 2014:
https://www.fas.org/sgp/crs/row/R42601.pdf

Lynch, Marc, *The Arab Uprising: The Unfinished Revolutions of the New Middle East*, New York: PublicAffairs, 2013.

MacKinnon, Rebecca, *Consent of the Networked: The Worldwide Struggle for Internet Freedom,* New York: Basic Books, 2012.

Manhken, Thomas G., ed., *Competitive Strategies for the 21st Century: Theory, History, and Practice*, Stanford, Calif.: Stanford University Press, 2012.

Morozov, Evgeny, *The Net Delusion: The Dark Side of Internet Freedom,* New York: PublicAffairs, 2011.

Office of the Secretary of Defense, *Annual Report to Congress: Military and Security Developments Involving the People's Republic of China 2013*, U.S. Department of Defense, 2013. As of February 24, 2014:
http://www.defense.gov/pubs/2013_china_report_final.pdf

Olson, Mancur, "The Logic of Collective Action in Soviet-Type Societies," *Journal of Soviet Nationalities*, Vol. 1, No. 2, Summer 1990.

Open Net Initative, *Profiles: China,* August 9, 2012. As of February 24, 2014: https://opennet.net/research/profiles/china

Page, Jeremy, "Internal Security Tops Military in China Spending," *Wall Street Journal*, March 5, 2011. As of February 24, 2014: http://blogs.wsj.com/chinarealtime/2011/03/05/internal-security-tops-military-in-china-spending/

Parker, Emily, *Now I Know Who My Comrades Are: Voices from the Internet Underground*, New York: Sarah Crichton Books, 2014.

Pettyjohn, Stacic L., "Net Gain: Washington Cedes Control Over ICANN," *Foreign Affairs*, April 10, 2014. As of July 7, 2014: http://www.foreignaffairs.com/articles/141122/stacie-l-pettyjohn/net-gain

Phys.org, "Factors Identified That Influence Willingness to Use Technology," March 8, 2013. As of February 24, 2014: http://phys.org/news/2013-03-factors-willingness-technology.html

Raub, Werner, and Chris Snijders, "Gains, Losses, and Cooperation in Social Dilemmas and Collective Action: The Effects of Risk Preferences," *Journal of Mathematical Sociology*, Vol. 22, No. 3, 1997.

Reporters Without Borders, *Enemies of the Internet 2013 Report: Special Edition Surveillance*, Paris, France, 2013.

Roshandel, Jalil, *Iran, Israel, and the United States: Regime Security vs. Political Legitimacy*, Santa Barbara, Calif.: Praeger Security International, 2011.

Schmidt, Eric, and Jared Cohen, "The Digital Disruption: Connectivity and the Diffusion of Power," *Foreign Affairs*, Vol. 80, No. 6, November/December 2010.

Shirky, Clay, "The Political Power of Social Media: Technology, the Public Sphere, and Political Change," *Foreign Affairs*, Vol. 90, No. 1, January/February 2011.

———, *Here Comes Everybody: The Power of Organizing Without Organizations*, New York: Penguin Books, 2011.

Silberglitt, Richard, and Lance Sherry, *A Decision Framework for Prioritizing Industrial Materials and Research and Development*, Santa Monica, Calif.: RAND Corporation, MR-1558-NREL, 2002. As of February 24, 2014: http://www.rand.org/pubs/monograph_reports/MR1558.html

Silberglitt, Richard, Lance Sherry, Carolyn Wong, Michael Tseng, Emile Ettedgui, Aaron Watts, and Geoffrey Stothard, *Portfolio Analysis and Management for Naval Research and Development*, Santa Monica, Calif.: RAND Corporation, MG-271-NAVY, 2004. As of February 24, 2014: http://www.rand.org/pubs/monographs/MG271.html

Sutter, Robert G., *Chinese Foreign Relations: Power and Policy Since the Cold War*, Lanham, Md.: Rowman and Littlefield Publishers, 2012.

Tkacheva, Olesya, Lowell H. Schwartz, Martin C. Libicki, Julie E. Taylor, Jeffrey Martini, and Caroline Baxter, *Internet Freedom and Political Space*, Santa Monica, Calif.: RAND Corporation, RR-295-DOS, 2013. As of February 24, 2014: http://www.rand.org/pubs/research_reports/RR295.html

U.S. Department of State, *Internet Freedom*, undated. As of February 24, 2014: http://www.state.gov/e/eb/cip/netfreedom/index.htm

Zickuhr, Kathryn, *Who's Not Online and Why*, Pew Research Center, September 25, 2013.

Zittrain, Jonathan, and John Palfey, "Internet Filtering: The Politics and Mechanisms of Control," in Ron Deibert, John Palfrey, Rafal Rohozinski, and Jonathan Zittrain, eds., *Access Denied: The Practice and Policy of Global Internet Filtering*, Cambridge, Mass.: MIT Press, 2008.

Zuckerman, Ethan, "Internet Freedom: Beyond Circumvention," *My Heart's in Accra* (blog), February 22, 2010. As of July 16, 2014: http://www.ethanzuckerman.com/blog/2010/02/22/internet-freedom-beyond-circumvention/